Superconductor/ Ferromagnet Nanostructures

An Illustration of the Physics of Hybrid Nanomaterials

Superconductor/ Ferromagnet Nanostructures

An Illustration of the Physics of Hybrid Nanomaterials

Oriol T Valls

University of Minnesota, USA

World Scientific

NEW JERSEY · LONDON · SINGAPORE · BEIJING · SHANGHAI · HONG KONG · TAIPEI · CHENNAI · TOKYO

Published by

World Scientific Publishing Co. Pte. Ltd.

5 Toh Tuck Link, Singapore 596224

USA office: 27 Warren Street, Suite 401-402, Hackensack, NJ 07601

UK office: 57 Shelton Street, Covent Garden, London WC2H 9HE

Library of Congress Cataloging-in-Publication Data

Names: Valls, Oriol, author.

Title: Superconductor/ferromagnet nanostructures : an illustration of the
   physics of hybrid nanomaterials / Oriol Valls.

Description: Hackensack, New Jersey : World Scientific, [2022] |
   Includes bibliographical references and index.

Identifiers: LCCN 2021049424 | ISBN 9789811249563 (hardcover) |
   ISBN 9789811249570 (ebook for institutions) | ISBN 9789811249587 (ebook for individuals)

Subjects: LCSH: Nanostructures. | Heterostructures. | Superconductors. |
   Ferromagnetic materials. | Transport theory.

Classification: LCC QC176.8.N35 V35 2022 | DDC 620.1/15--dc23/eng/20211208

LC record available at https://lccn.loc.gov/2021049424

British Library Cataloguing-in-Publication Data

A catalogue record for this book is available from the British Library.

For any available supplementary material, please visit
https://www.worldscientific.com/worldscibooks/10.1142/12649#t=suppl

Desk Editor: Joseph Ang

Typeset by Stallion Press
Email: enquiries@stallionpress.com

To Maureen

Preface

Modern and ever-improving fabrication techniques have made feasible devices that not long ago were mere theoretical dreams. Among these are layered heterostructures comprising magnetic and superconducting layers, with well-defined interfaces. When desired, the layers can be on the order of a few nanometers thin.

The existence of such nanostructures has made it possible to experimentally study, and exploit the consequences of, the ferromagnet/superconductor proximity effects, which are in their primary form, short range and require thin layers and well-defined interfaces. To the intellectual and theoretical importance of understanding the intricacies of these phenomena is added the possible technological significance of these heterostructures for possible uses in spintronics in general, nonvolatile memory elements, and other device applications.

The book is intended for readers who want to get started in understanding the physics of hybrid ferromagnet/superconductor nanostructures. Therefore, it has a rather introductory and, to some extent, elementary character. It assumes only that the reader has a basic understanding of superconductivity, in particular the Bardeen-Cooper-Schrieffer (BCS) theory, and of magnetism at the level of the Hubbard model. Such topics are covered in introductory solid state textbooks. Concepts and theoretical developments beyond that are introduced in the text. Any person with a background in condensed matter physics at the level of a beginning graduate student should, I hope, be able to benefit from reading the material covered here and be able to read the more advanced literature and be ready to start contributing to the field.

The scope of the book is, therefore, relatively limited. It covers the basics of the subject: the purpose is to enable the reader to go on to assimilate the current and more advanced literature. The emphasis is very heavily theoretical, although one complete chapter, and some additional sections, is devoted to discussions of and comparisons with experiments. The clean limit is very heavily emphasized: this is because sample fabrication is constantly improving; samples are progressively getting better. The results are of course much more reproducible. Clean limit results on good samples are also much easier to interpret. Pedagogically, the clean limit results are easier to explain and to interpret. Clean does not mean perfect, however, and the effects of unavoidable surface imperfections and mismatches are properly and extensively discussed.

A considerable part of the material in this book consists of pedagogically rewritten and rearranged versions of work performed in my group over the years. This body of work could not have been achieved without the effort and initiative of the students that worked on it: Klaus Halterman, Paul Barsic, Chien-Te Wu, and Evan Moen. I must credit Prof. Igor Žutić, a former student of mine, with having gotten me interested in the subject. Experimental collaborators have provided me with invaluable insights on how samples are really made and how measurements are really taken: I particularly thank Prof. Ilya Krivorotov, Dr. Alejandro Jara and Dr. Francesca Chiodi.

Oriol T. Valls

Contents

Chapter 1

Introduction

1.1 Structures and Materials Considered

This book is about nanostructures comprising Superconducting materials (denoted as S), Ferromagnets (F) and metals that are neither superconducting nor magnetic, which we will denote as N. The superconductors that will be considered are ordinary, s-wave materials well-described by the Bardeen-Cooper-Schrieffer (BCS) theory. Similarly, we will describe the magnets via a simple effective field model. That is, the emphasis, indeed the subject of the book, is the effects that result from the interaction between the different kinds of order: magnetic and superconducting.

We will nearly exclusively deal with layered structures, with the layers being large in the transverse direction, while their thicknesses are, on the other hand, sufficiently small so that the important proximity effects due to the competing electronic orders are prominent. This is of course the important situation.

1.2 The Proximity Effect

Superconductivity is known to be, so to speak, contagious. If a normal (non-magnetic, non-superconducting) metal is put in contact with an ordinary superconductor, the Cooper pairs penetrate into this N material for a considerable distance. This has been known for a long time. Indeed, it can be easily shown (see e.g., Ref. [1]) that if one assumes that the Cooper pair amplitude decays exponentially with distance over a characteristic proximity length ξ_p then formally ξ_p is inversely proportional to the absolute temperature T. This of course does not mean that if one puts a meter-thick copper layer on top of, say, aluminum at very low T the Cu will

1

become superconducting. What it means is that the decay as $T \to 0$ is a power law, not an exponential. Basically what happens is that, in N materials, although there is no attraction (phonon mediated or not) that causes Cooper pairs to form, there is usually no repulsion either that may cause Cooper pairs created elsewhere to break up. So, they live happily, not ever after, but for quite a while. We speak of the proximity effect being long-ranged.

It is quite another matter when the material in contact with the superconductor is ferromagnetic. Intuitively speaking, the internal magnetic field in the F material will break the singlet Cooper pairs. This will be a strong effect, which will result in the proximity effect being short-ranged. This intuition is basically correct, but there are two wrinkles: one is that the penetration of the Cooper pair amplitude is oscillatory. We will see how this occurs and why in Chap. 2. The existence of these oscillations leads to a variety of nonmonotonic behaviors, as we shall see, which is one of the things that makes these systems particularly interesting. The second is even more important: the same intuition that tells us that the internal field will break the singlet Cooper pairs ($S = 0$ and $S_z = 0$ being their spin state) tells us that this does not have to be the case with $S = 1$ and $S_z = \pm 1$ pair amplitudes. Since we are dealing with s-wave superconductors, the issue is whether singlet to triplet pairing conversion is possible. I shall discuss this in Chap. 3 where we will see that the answer is that, under the right circumstances, such a conversion does occur. When it does, it must be by converting the Cooper pairs to a rather peculiar state of triplet amplitudes. This is due to fundamental constraints arising from conservation laws and the Pauli principle.

There is also a *reverse* proximity effect: the magnetization will penetrate into the S material and also will be somewhat depleted in the F layer or layers, near the interfaces. Although this reverse effect is less important, it is nontrivial and we will also discuss it.

1.3   What the Book Covers

Thus, this book is about the interaction between materials in close proximity whose properties are well-understood. One can of course study proximity effects in exotic materials, i.e., if the superconductor or the magnetic state are exotic. Such complications will at most be briefly touched upon.

The level of the book is such that I expect it to be readable with profit by anybody familiar with basic Solid State physics, including standard BCS theory and elementary magnetism. Ownership of any one of the classical books on BCS superconductivity, particularly the book by DeGennes [2] or that by Tinkham [3] will be found invaluable. My objective is that after reading this book, or relevant parts of it, such a person is ready to tackle all of the literature on the subject.

This book will deal with the case where the *bulk* materials that are part of the samples are reasonably clean, as they are for modern well-fabricated structures. The *interfaces* between materials are of course another story. Interfacial scattering is unavoidable; they can never be assumed to be perfect, but they will be assumed to be good enough so that the relevant proximity effects exist and can be experimentally studied. After all, if the interfaces are bad enough there are hardly any proximity effects that are not exponentially small. As we will see, the parameter ranges for interfacial scattering (as well as material and geometrical values) that we will consider are well within the capabilities afforded by current fabrication techniques.

Proximity effects are at the core of all that is done in the book. This means that it will consider nanostructure configurations in which typical material thicknesses are of order of, or smaller than, the largest of the relevant microscopic lengths and usually in the nanoscale range. Size effects are fundamental and the finite thicknesses of the layers become important geometric lengths in the proximity effects. Therefore we will tackle the problem using methods that do not involve coarse graining over atomic length scales, as would be the case with quasiclassical [4] methods that may not be appropriate when the thickness of the materials is only of a few atomic layers.

1.4 Organization

After discussing the basic formalism in Chapter 2, the book will move on to a complete discussion of conservation laws, singlet to triplet conversion, and the calculation of thermodynamic properties. This culminates in a detailed comparison with experiment. Afterwards, in the second part of the book, the calculation of transport properties will be presented. We will see that in these structures it is important to consider not only charge, but also spin transport: important spin transfer properties are present, which make spin manipulation and consequent applications to spintronics possible.

Although the basic equations are fairly easy to derive numerically, as we will see, obtaining final results involves repeated numerical diagonalization of large matrices. Therefore final results are usually given in terms of organized sets of figures, of which the book has a very large number.

Chapter 2

The Basic Formalism

2.1 Introduction

In this chapter the main formalism that will subsequently be used, in several variations throughout the book, is developed and considered. Adequate space will be devoted not only on the basic theoretical structure but also on analytical and numerical considerations, so that details and techniques discussed here will not need to be repeated later on.

The procedure is in principle very simple: since under the assumptions that will be made the Hamiltonian is known, one can just go ahead and diagonalize it. Once the eigenvalues and eigenfunctions are known, any relevant physical quantity can be calculated. These are purely microscopic methods. As explained in Chap. 1 no averaging over quantum length scales will be performed. Such averaging would introduce serious weaknesses since we will be dealing with sharp, atomic scale interfaces and in many cases, nanoscale (or below) layer thicknesses.

There are however a few complications: the first one is that the effective Hamiltonian is known only in terms of the effective BCS pair potential $\Delta(\mathbf{r})$ which must be determined self-consistently. This means that one has to diagonalize time and again, iteratively, until self-consistency is reached. We will see that this is quite achievable, to great precision, but we will also see that it must be taken care of or very bad things (such as conservation law violations) are likely to happen. The second is that the diagonalization must be performed numerically. This turns out to be not very difficult in practice but the details, and how long it takes, depend on the computer power available.

As mentioned in the second paragraph of Chap. 1, this book will focus on layered systems of largely nanoscale layer thicknesses. Basic sketches

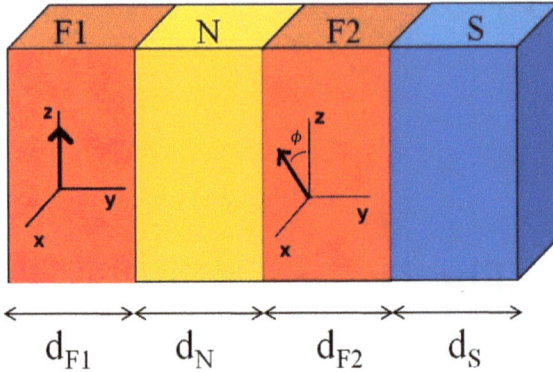

Fig. 2.1 Schematic of a layered system in the so-called spin valve configuration. The notation for the thicknesses is indicated. The diagram is not to a realistic scale. The direction of the internal Stoner fields (see text) is indicated.

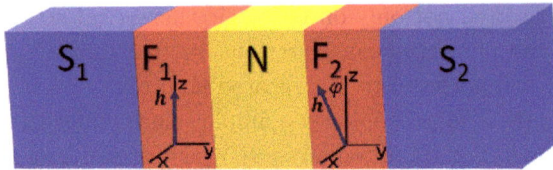

Fig. 2.2 Schematics (not to scale) of a layered structure of the ferromagnetic Josephson junction type.

of generic but important examples of the layered structures we will study throughout the paper are in Figs. 2.1 and 2.2. Simpler structures will also be studied, largely for pedagogical purposes. More complicated structures will be displayed later. The two examples shown in these two figures are what we will call the spin valve and Josephson types respectively.

For layered structures, as we shall see below, the necessary calculations can be performed independently for each value of the transverse energy. This makes it natural to use parallel processing: every transverse mode can be diagonalized independently, being sent to a different processor. Only after all processors are done with their individual tasks must one put everything together in the self-consistency condition. This condition involves, a sum over *all* relevant energies. In practice, therefore, the procedure turns out to be very fast and efficient even in only moderately powerful computer systems. See Sec. 2.5.2 for details.

2.2 Materials: Considerations

Standard, well understood BCS superconductors are assumed. Many of the comparisons with experiment will specifically consider S layers or layers made of Niobium. In these comparisons the F material considered is Cobalt, but comparisons with the more complicated Holmium structure will also be made in Chap. 5. The magnetism will be treated via a simple effective field model. The materials we consider in the book are assumed to be in the clean limit, although no such assumption is made for the interfaces separating them, which are assumed to be realistically good samples with unavoidable surface roughness. It is possible to make very clean bulk materials, but not perfect interfaces.

2.3 The Hamiltonian

As just mentioned, the assumptions made about the simple nature of the materials involved imply that the *effective* Hamiltonian of our system, $\mathcal{H}_{\text{eff}}$, is known. It can be very generically written as:

$$
\mathcal{H}_{\text{eff}} = \int d^3r \left\{ \sum_{\alpha} \psi_{\alpha}^{\dagger}(\mathbf{r}) \mathcal{H}_e \psi_{\alpha}(\mathbf{r}) + \frac{1}{2} \left[\sum_{\alpha,\beta} (i\sigma_y)_{\alpha\beta} \Delta(\mathbf{r}) \psi_{\alpha}^{\dagger}(\mathbf{r}) \psi_{\beta}^{\dagger}(\mathbf{r}) + \text{h.c.} \right] \right.
$$
$$
\left. - \sum_{\alpha,\beta} \psi_{\alpha}^{\dagger}(\mathbf{r}) (\mathbf{h} \cdot \boldsymbol{\sigma})_{\alpha\beta} \, \psi_{\beta}(\mathbf{r}) \right\}, \tag{2.1}
$$

where

$$
\mathcal{H}_e = -1/(2m)\boldsymbol{\nabla}^2 - E_F + U(\mathbf{r}). \tag{2.2}
$$

In the notation we will use throughout the book $\boldsymbol{\sigma}$ are the set of Pauli matrices, with spin denoted by Greek indices. The operators $\psi_{\sigma}(\mathbf{r})$ (and $\psi_{\sigma}^{\dagger}(\mathbf{r})$) are the usual annihilation (and creation) operators. We represent the magnetism of the F layers by an effective exchange field, called the Stoner field, which produces a spin-dependent energy shift. In general $\mathbf{h}(\mathbf{r})$ will have components in all directions. When there are several F materials involved, this field may be different in each one of them. In general $\mathbf{h}(\mathbf{r})$ vanishes in N and S.

The quantity $\Delta(\mathbf{r})$ is the BCS singlet pair potential, to be calculated self-consistently as discussed below:

$$
\Delta(\mathbf{r}) \equiv gF(\mathbf{r}) \tag{2.3}
$$

where g is the usual singlet BCS coupling constant, nonvanishing in the S layers, and the quantity

$$F(\mathbf{r}) \equiv \langle \psi_\uparrow(\mathbf{r})\psi_\downarrow(\mathbf{r}) \rangle \qquad (2.4)$$

is the *pair amplitude*. In inhomogeneous systems one should be particularly careful not to confuse it with the pair potential. Only the self-consistent potential leads to a solution (Ref. [2]) corresponding to a local minimum of the free energy. This reflects that the Hamiltonian 2.1 is an effective Hamiltonian which represents the original system only when the pair potential is evaluated self-consistently according to Eq. (2.3).

In Eq. (2.2) m is an effective mass, E_F is the relevant bandwidth, i.e., the energy difference between the bottom of the band and the chemical potential. We denote its value in S by E_{FS}: it will in general be different in the F material (or materials). There we can call (see Fig. 2.3) its mean value E_{FM}, so that in the proper units as discussed below, the bandwidths of the up and down magnetic bands are $E_{FM} \pm h$. A crude but informative scheme is shown in Fig. 2.3. The zero of energy will always be taken at E_F. The quantity $U(\mathbf{r})$ represents the interfacial scattering potential arising from unavoidable imperfections in the sample growth process. This potential can be taken to be nonvanishing, for good samples, except very near the interfaces, and a delta function approximation will be used in this book.

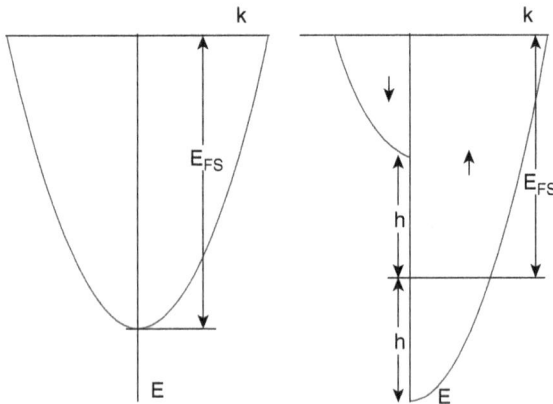

Fig. 2.3 Rough illustrative scheme of the different bandwidths defined.

2.4 The Bogoliubov–DeGennes Equations

To diagonalize the effective Hamiltonian, the field operators $\psi_\alpha^\dagger$ and ψ_α are rewritten by means of a Bogoliubov transformation, which, choosing a common phase convention [5], we write as:

$$\psi_\uparrow(\mathbf{r}) = \sum_n \left(u_{n\uparrow}(\mathbf{r})\gamma_n - v_{n\uparrow}(\mathbf{r})\gamma_n^\dagger \right), \tag{2.5a}$$

$$\psi_\downarrow(\mathbf{r}) = \sum_n \left(u_{n\downarrow}(\mathbf{r})\gamma_n + v_{n\downarrow}(\mathbf{r})\gamma_n^\dagger \right), \tag{2.5b}$$

where $u_{n\alpha}$ and $v_{n\alpha}$ are the spin-dependent quasiparticle and quasihole amplitudes, and γ_n and $\gamma_n^\dagger$ are the fermionic Bogoliubov quasiparticle annihilation and creation operators, respectively. The sum is over an appropriate set (see below) of quantum numbers. The equations that result from this transformation are the Bogoliubov-DeGennes (see Ref. [2]) equations. We require that the transformations in Eq. (2.5) diagonalize $\mathcal{H}_{\mathrm{eff}}$,

$$[\mathcal{H}_{\mathrm{eff}}, \gamma_n] = -\epsilon_n \gamma_n, \tag{2.6a}$$

$$[\mathcal{H}_{\mathrm{eff}}, \gamma_n^\dagger] = \epsilon_n \gamma_n^\dagger. \tag{2.6b}$$

One takes the commutators $[\psi_\alpha(\mathbf{r}), \mathcal{H}_{\mathrm{eff}}]$ and uses the standard Fermion commutation relations to obtain expressions for the commutators in Eq. (2.6).

The above is quite general. We can now begin to specialize to the layered geometry. The system is assumed to be very large in the plane perpendicular to the layers, which I label as the $x - z$ plane, while the layer thicknesses are finite (in practice nanoscale). The magnetizations of the F layers are usually, in this plane, due to demagnetization effects. In this basic formalism section I will not include any normal to the layers magnetization component, although later (in Chap. 11 and Sec. 5.3 for example) I will add such a component. All $\mathbf{r}$ dependent parameters, such as the pair potential and the internal fields, as well as the interfacial scattering, depend therefore only on the y coordinate. We have then the commutators,

$$[\psi_\uparrow(\mathbf{r}), \mathcal{H}_{\mathrm{eff}}] = (\mathcal{H}_e - h_z)\psi_\uparrow(\mathbf{r}) - h_x\psi_\downarrow(\mathbf{r}) + \Delta(\mathbf{r})\psi_\downarrow^\dagger(\mathbf{r}), \tag{2.7a}$$

$$[\psi_\downarrow(\mathbf{r}), \mathcal{H}_{\mathrm{eff}}] = (\mathcal{H}_e + h_z)\psi_\downarrow(\mathbf{r}) - h_x\psi_\uparrow(\mathbf{r}) - \Delta(\mathbf{r})\psi_\uparrow^\dagger(\mathbf{r}). \tag{2.7b}$$

Inserting (2.5) into (2.7) and using Eq. (2.6) yields the general spin-dependent BdG equations,

$$\begin{pmatrix} \mathcal{H}_0 - h_z(y) & -h_x(y) & 0 & \Delta(y) \\ -h_x(y) & \mathcal{H}_0 + h_z(y) & \Delta(y) & 0 \\ 0 & \Delta(y) & -(\mathcal{H}_0 - h_z(y)) & -h_x(y) \\ \Delta(y) & 0 & -h_x(y) & -(\mathcal{H}_0 + h_z(y)) \end{pmatrix}$$

$$\times \begin{pmatrix} u_{n\uparrow}(y) \\ u_{n\downarrow}(y) \\ v_{n\uparrow}(y) \\ v_{n\downarrow}(y) \end{pmatrix} = \epsilon_n \begin{pmatrix} u_{n\uparrow}(y) \\ u_{n\downarrow}(y) \\ v_{n\uparrow}(y) \\ v_{n\downarrow}(y) \end{pmatrix}, \qquad (2.8)$$

where the single particle Hamiltonian $\mathcal{H}_0$ is defined as,

$$\mathcal{H}_0 \equiv \frac{\widehat{p}_y^2}{2m} + \varepsilon_\perp - E_F + U(y). \qquad (2.9)$$

A plane wave factor $e^{i\mathbf{k}_\perp \cdot \mathbf{r}}$ has been canceled in both sides of Eq. (2.8). The longitudinal momentum operator, $\widehat{p}_y$, is given by, $\widehat{p}_y = -i\partial/\partial y$, $\varepsilon_\perp \equiv k_\perp^2/2m$ is the kinetic energy of the transverse modes and $\Delta(y)$ is the pair potential, to be evaluated self-consistently. The components of the internal magnetic fields **h** are nonvanishing only in the magnetic layer or layers. Whenever there are more than one magnetic layer, the direction and even the magnitude of these fields will in general differ from layer to layer, so it is understood that the h_x and h_z components in Eq. (2.8) are stepwise functions of y. Within our assumptions, the interfacial scattering potential consists of delta function terms at each interface. If there are $n_l + 1$ layers there will be n_l interfaces:

$$U(y) = \sum_{i=1}^{n_l} H_i \delta(y - d_i) \qquad (2.10)$$

where d_i are the interface locations and H_i strength parameters. We will find it convenient to express these parameters in dimensionless form as $H_{Bi} \equiv H_i/v_{FS}$. One can easily add spin dependence to these scattering strengths, whenever the interfaces are spin-active, as we will do in Sec. 5.2. The limit $H_B = 0$ corresponds to an unrealistic perfect interface while $H_B \gg 1$ represents an undesirable "tunneling limit" interface, which would kill all the proximity effects and all interesting behaviors.

Note that it is possible, if desired, to write many of the above matrix equations in a more compact way. If one introduces $\boldsymbol{\rho}$ as a set of Pauli-like matrices in particle-hole space in conjunction with the ordinary Pauli matrices σ in spin space one can rewrite Eqs. (2.8) in the more compact but perhaps less transparent way:

$$\left[\rho_z \otimes \left(\mathcal{H}_0\hat{1} - h_z\sigma_z\right) + \left(\Delta(y)\rho_x - h_x\hat{1}\right) \otimes \sigma_x\right] \Phi_n = \epsilon_n\Phi_n, \qquad (2.11)$$

where $\Phi_n \equiv (u_{n\uparrow}(y), u_{n\downarrow}(y), v_{n\uparrow}(y), v_{n\downarrow}(y))^{\mathrm{T}}$, with the T superscript denoting transposition.

2.5 Self-Consistency

As mentioned above, the pair potential in the effective Hamiltonian is a quantity that must be calculated self-consistently, since in the original derivation of the *effective* Hamiltonian, $\mathcal{H}_{\mathrm{eff}}$, it is related to the anomalous averages involving the $\psi(\mathbf{r})$ operators. Rewriting the ψ and $\psi^\dagger$ operators in terms of the Bogoliubov operators one obtains, via Eqs. 2.3 and 2.4 the self-consistency condition which relates the spectrum obtained from Eq. (2.8) to the inhomogeneous pair potential $\Delta(y)$ by an appropriate sum over states:

$$\Delta(y) = \frac{g(y)}{2}\sum_n{}' \left[u_n^\uparrow(y)v_n^\downarrow(y) + u_n^\downarrow(y)v_n^\uparrow(y)\right] \tanh(\epsilon_n/2T), \qquad (2.12)$$

where T is the temperature. The sum is over all eigenstates: it encompasses a sum over the continuous transverse energy $\varepsilon_\perp$ and the "longitudinal" index n,

$$\sum_n{}' \to \sum_{\varepsilon_\perp}{}'\sum_n{}'. \qquad (2.13)$$

The prime on the sum indicates that only those positive energy states with energy less than the pairing interaction energy cutoff, ω_D, are included. The function $g(y)$ vanishes in the F layers while in the S layers it takes the value of the usual BCS *singlet* coupling constant in the S material.

2.5.1 *Diagonalization procedures: Practicalities*

To diagonalize Eq. (2.8), that is, to find the *exact* eigenvalues and eigenfunctions of the system, we note first that these are a separate set of equations, one for each value of $\varepsilon_\perp$. Thus, the original diagonalization problem is, for the quasi-one dimensional geometry, of block form. The energy quantum numbers are given by the set $\varepsilon_\perp$ and n. To perform the diagonalization, we expand the wavefunctions in terms of any convenient complete set of eigenfunctions $\phi_q(y)$ and write:

$$u_n^\alpha(y) = \sum_{q=1}^{N} u_{nq}^\alpha \phi_q(y), \tag{2.14a}$$

$$v_n^\alpha(y) = \sum_{q=1}^{N} v_{nq}^\alpha \phi_q(y) \tag{2.14b}$$

for $\alpha = \uparrow, \downarrow$. As long as no current is flowing one can simply use sine functions:

$$\phi_q(y) = \sqrt{\frac{2}{d}} \sin(q\pi y/d) \tag{2.15}$$

with d being the *total* thickness of the sample. Of course, when a current is present one may have to use exponentials instead of sines, and periodic boundary conditions, as discussed later in the book. The use of complex functions may then be required. Similarly the pair potential can be expanded:

$$D_{qq'} = \int_0^d dy\, \phi_q(z)\Delta(y)\phi_{q'}(z), \tag{2.16}$$

where the integral is over the superconducting regions. The self-consistency condition Eq. (2.12) is now transformed into,

$$\Delta(y) = \frac{g(y)}{2} \sum_{p,p'} \sideset{}{'}\sum_{n} \left[u_{np}^\uparrow v_{np'}^\downarrow \phi_p(y)\phi_{p'}(z) + u_{np}^\downarrow v_{np'}^\uparrow \phi_p(y)\phi_{p'}(z) \right]$$

$$\times \tanh(\epsilon_n/2T). \tag{2.17}$$

It is then straightforward to rewrite Eq. (2.8) as a matrix equation involving matrix elements with respect to the indices q. The matrix elements of the Hamiltonian with respect to these indices are simply obtained by multiplying the different terms by the appropriate basis functions and integrating over y. Similarly, the matrix elements $D_{q,q'}$ of $\Delta(y)$ can be evaluated. The results depend on the geometry considered (number and thickness of the layers) but they are always straightforward, if tedious, to compute. Because of this, I will give only one example here, and leave those needed in the future as exercises.

The example I will give here is that of an F_1SF_2 system (see Fig. 3.1 in the next Chapter). I assume a uniform $E_F = E_{FS} = E_{FM}$ (Fig. 2.3) and identical scattering at the two interfaces. The two internal fields are oriented at angles $\pm\alpha/2$ with respect to the z axis. I give the matrix elements in dimensionless form in terms of E_F. In the Eq. (2.15) basis, all the integrals are elementary. Writing the matrix involved in Eq. (2.8) in the shorthand:

$$
\begin{bmatrix}
H_0 - H_z & -H_x & 0 & D \\
-H_x & H_0 + H_z & D & 0 \\
0 & D & -(H_0 - H_z) & -H_x \\
D & 0 & -H_x & -(H_0 + H_z)
\end{bmatrix}
\tag{2.18}
$$

one gets the results, after some work:

$$
(H_0)_{mn} = \left[\left(\frac{m\pi}{k_F d} \right)^2 \frac{\varepsilon_\perp}{E_F} - 1 \right] \delta_{mn} + Z_B [\mathcal{U}_{m-n}(d_{F1})
$$
$$
+ \mathcal{U}_{m-n}(d_{F1} + d_S) - \mathcal{U}_{m+n}(d_{F1}) - \mathcal{U}_{m+n}(d_{F1} + d_S)],
\tag{2.19a}
$$

$$
(H_z)_{mn} = \frac{h_0}{E_F} \cos(\alpha/2) \big[\mathcal{K}_{m-n}(d_{F1}) - \mathcal{K}_{m+n}(d_{F1}) + \mathcal{K}_{m+n}(d_{F1} + d_S)
$$
$$
- \mathcal{K}_{m-n}(d_{F1} + d_S) \big],
$$

$(form \neq n),$
$\tag{2.19b}$

$$
= \frac{h_0}{E_F} \cos(\alpha/2) \left[\frac{d_{F1} + d_{F2}}{d} + \mathcal{K}_{2m}(d_{F1} + d_S) - \mathcal{K}_{2m}(d_{F1}) \right],
$$

$(form = n),$
$\tag{2.19c}$

$$
(H_x)_{mn} - \frac{h_0}{E_F} \sin(\alpha/2) \big[\mathcal{K}_{m+n}(d_{F1}) - \mathcal{K}_{m-n}(d_{F1}) + \mathcal{K}_{m+n}(d_{F1} + d_S)
$$
$$
- \mathcal{K}_{m-n}(d_{F1} + d_S) \big], \quad m \neq n,
\tag{2.19d}
$$

$$
= \frac{h_0}{E_F} \sin(\alpha/2) \left[\frac{d_{F2} - d_{F1}}{d} + \mathcal{K}_{2m}(d_{F1} + d_S) + \mathcal{K}_{2m}(d_{F1}) \right],
$$

$(form = n),$
$\tag{2.19e}$

$$
(D)_{mn} = \frac{2}{E_F d} \int_{d_{F1}}^{d_{F1} + d_S} dy \, \sin \left[\frac{m\pi y}{d} \right] \Delta(y) \sin \left[\frac{n\pi y}{d} \right],
\tag{2.19f}
$$

where $Z_B \equiv 2H_B/(k_F d)$. We have also defined:

$$
\mathcal{K}_n(y) \equiv \frac{\sin\left(\frac{n\pi y}{d}\right)}{n\pi}, \quad \mathcal{U}_n(y) \equiv \cos\left(\frac{n\pi y}{d}\right).
\tag{2.20}
$$

2.5.2 *Diagonalization procedures: Numerical considerations*

The diagonalization involves of course very large matrices. Formally, indeed, the number of values N_q of the indices q as well as that of the $\varepsilon_\perp$, $N_\perp$, is infinite. In practice, however, we are concerned with relevant energies that have a maximum value, at the most, of a few times ω_D above or below the chemical potential. This determines the maximum needed values of q and of $\varepsilon_\perp$. For the first, the maximum would be of order $k_F d/\pi$ (with d being the total sample thickness), plus a correction of order ω_D/E_F. For typical samples (see e.g., Chap. 6) this would amount to a few hundred or up to a thousand. For the $\varepsilon_\perp$ one must take a sufficiently fine mesh between zero and the maximum. In practice this turns out to be between 500 and 5000 depending on the problem. Since the diagonalization for each $\varepsilon_\perp$ value is independent of that for other values, the matrices that one must diagonalize are not forbiddingly large (size $N_q \times N_q$) although the diagonalization must be performed for a large number ($N_\perp$) of matrices. Efficient canned routines for diagonalizing matrices are available in any computer system. For best clock time performance, the use of parallel processing is very highly recommended: matrices corresponding to different values of $\varepsilon_\perp$ are simply diagonalized by different processors, and the MPI (Message Passing Interface) or similar amount of programming needed when putting the results together is truly minimal. As in any numerical problem, a certain amount of trial and error is required to find out the optimal cutoffs and mesh sizes, but the efficiency of the parallelized computations makes it easy to err on the conservative size.

As mentioned before, self-consistency is essential and requires repeated diagonalization until Eq. (2.12) is satisfied. To start the process, one makes an initial guess for the shape of $\Delta(y)$, usually based on square waves: constant values in S and zeros in N or F_i. Even better, when one is solving for a set of parameter values only slightly different from a set one has already solved for before then one uses the self-consistent solution of the previous set as the starting point for the next. How large a number of iterations one needs depends on the quality of the guess: it can be half a dozen or less for a good one, but when the system transitions to a different state (see below, Chap. 4) then the number can be up to eight or ten times larger. Also, one must set the criterion for convergence: as usual one must not set it too loose (which one can tell because the results change on making it more strict) nor unnecessarily tight, which wastes resources. Experience is the best guide and good judgment is needed, but the art is easily mastered.

2.6 Parameters and Normalizations

Throughout the book, we will most of the time use dimensionless quantities normalized in a standard way. The magnets will be characterized by their dimensionless internal field, redefined as $h \to h_0/E_{FM}$, where here I denote the original internal field energy by h_0. In the limit $h = 1$ the system is half metallic (see Fig. 2.3). In general, it is convenient to normalize all other energies (including the eigenenergies) by E_{FS}, except the pair potential, which is best normalized by its bulk value in S, denoted as Δ_0. If there are several different magnetic materials involved then there will be several such fields. Lengths will usually be given in dimensionless form, multiplied by k_{FS} and denoted by a capital letter, e.g., $Y \equiv k_{FS}y$. As to the S material, it is characterized by its coherence length ξ_0. By a textbook BCS relation $\Xi_0 = (2/\pi)E_{FS}/\Delta_0$. Temperatures are best normalized via T_c^0, the transition temperature of bulk S. The only additional thing that needs to be specified is ω_D but the normalized results are, not surprisingly, independent of ω_d/E_{FS} provided that this quantity is small, as it should be. This is similar to what occurs in bulk BCS calculations where it is related to quantities such as T_c/Δ being independent of ω_D. In addition, since in general E_{FS} and E_{FM} are not the same, a mismatch parameter $\Lambda \equiv (E_{FM}/E_{FS})$ (or more than one if different materials are involved) must be introduced. As mentioned below Eq. (2.10), the interfacial scattering parameters are conveniently made dimensionless by dividing them by v_{FS}, the Fermi velocity in S. Finally, in transport calculations the conductance is normalized to the conductance quantum $G_q = e^2/2\pi\hbar$. These conventions are summarized in Table 2.1.

Table 2.1 Dimensionless variables used.

Physical Quantity	Dimensionless Form
Exchange energies	$h \equiv h_0/E_{FM}$
Fermi surface mismatch	$\Lambda \equiv (E_{FM}/E_{FS})$
Temperature	$t \equiv T/T_c^0$
Coherence length	$\Xi_0 \equiv k_{FS}\xi_0$
Debye energy	$\omega \equiv \omega_D/E_{FS}$
Barrier strengths	$H_{Bi} \equiv H_i/v_{FS}$
Distances generically	$Y \equiv k_{FS}y$
Dimensionless time	$\tau \equiv \omega_D t$
Conductance	$G \equiv G/G_q$

Among the parameters listed in the table, we shall see that the geometrical parameters, as described by the dimensionless lengths, are always highly relevant. Of the physical quantities, the dimensionless exchange energies h are always very important, as are the surface scattering strengths H_{Bi}. The dependence of the results on these quantities will be discussed in detail. The effect of the mismatch parameter Λ is to produce interfacial scattering and this can to *some* extent be subsumed with the barriers. Whenever results are shown without indicating its value, the reader should assume that it has been set to unity.

2.7   Pair Amplitude and Characteristic Lengths

It is convenient to introduce here characteristic lengths that describe the proximity effect. This introduction can be most simply performed by considering the simplest case, a two-layer system where only one S and one F layer are present. Some results for the pair amplitude $F(Y)$ (see Eq. (2.4)) in that case are shown in Fig. 2.4 (from Ref. [6]). The pair amplitude is shown in the magnetic and superconductor sides in the top and bottom panels respectively. Results for a wide range of temperatures are shown. The magnet is characterized by an intermediate value $h = 0.5$. $F(Y)$ is normalized to its zero T bulk value $F_0 = \Delta_0/g$. The two regions $Y > 0$ (superconductor) and $Y < 0$ (normal) are plotted in separate panels because their significant features occur over different horizontal and vertical scales. The pair amplitude, however, as opposed to the pair potential, is continuous across the interface. On the magnetic side, $F(Y)$ has an oscillatory behavior, very different from the monotonic behavior at $h = 0$. This is because at $h \neq 0$ the spin degeneracy is removed. The Fermi wave vectors of the spin up and spin down electrons, $k_{F\uparrow}, k_{F\downarrow}$, are different, and consequently a Cooper pair entering the ferromagnet acquires a net center of mass momentum. The superposition of the spin up and down wavefunctions causes the superconducting wavefunction to oscillate on a length scale set by the difference in the spin up and spin down wavevectors. One can define a length xi_2 via $\xi_2 = (k_{F\uparrow} - k_{F\downarrow})^{-1}$. We have

$$k_{FS}\xi_2 = \frac{1}{(\Lambda(1+h))^{1/2}} \frac{k_{F\uparrow}}{k_{F\uparrow} - k_{F\downarrow}} \qquad (2.21)$$

where Λ is the wavevector mismatch parameter of Table 2.1. Since for h not too large $k_{F\uparrow}\xi_2 \approx 1/h$, the characteristic length of oscillations scales

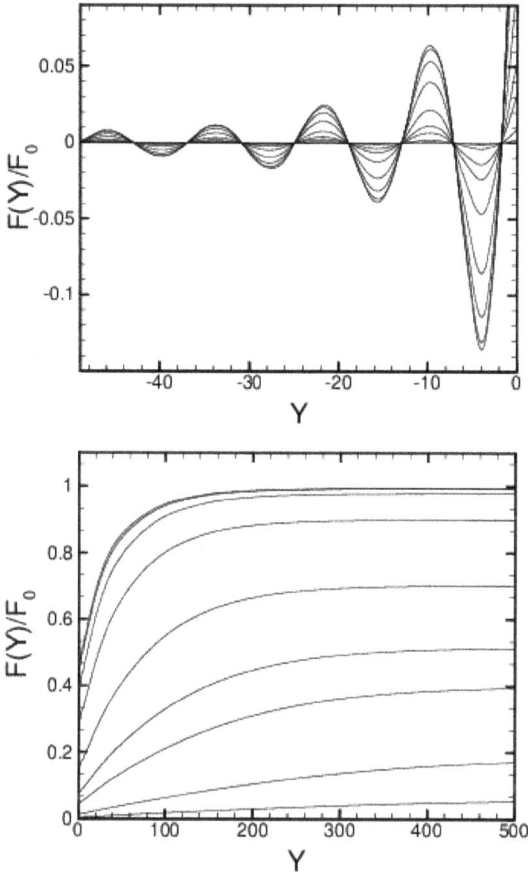

Fig. 2.4 Example of the pair amplitude $F(Y)$, normalized to its zero T bulk value in S, plotted as a function of dimensionless distance $Y = k_{FS}z$ from a F/S interface. Results are for $h = 1/2$. The top panel depicts the magnetic region, while the bottom panel shows the superconducting region. The curves correspond, from top to bottom on the superconducting side or in order of decreasing amplitude in the magnetic side, to temperatures $t \equiv T/T_c = 0, 0.2, 0.4, 0.6, 0.8, 0.9, 0.94, 0.98, 0.99$. Note the different vertical and horizontal scales in both panels.

approximately as $1/h$, and therefore, except at extremely small h, it is much smaller than length scale set in the normal metal case above.

For most exchange fields, and except extremely near the the interface, the pair amplitude is given approximately by

$$F(Y) = \alpha \frac{\sin[Y/(k_{FS}\xi_2)]}{Y/(k_{FS}\xi_2)}, \qquad T = 0, \qquad (2.22)$$

where α is a constant. Very close to the interface, the pair amplitude monotonically decays over a characteristic length ξ_1 which can be defined as the
first point inside the ferromagnet at which $F(Z)$ is zero. The length scale
ξ_1 scales like xi_2 as $k_{FS}\xi_1 \approx 1/h$.

Starting with Eq. (2.22) we can see, as T increases at finite h, how the
behavior of the pair amplitude near the interface changes. Starting with
the largest amplitude curve in the upper panel of Fig. 2.4 ($T = 0$), we see
that, beyond a small region of fast decay at the interface, the pair amplitude
exhibits damped oscillations, with a temperature independent period that
coincides with the expected value $k_{FS}\xi_1 \approx 1/h = 2$, independent of T. The
envelope decay of the oscillations varies inversely with distance as given in
Eq. (2.22). The quantity ξ_1 is also independent of temperature, since as can
be seen in the figure the location of the first node of $F(Y)$ as it monotonically goes to zero near the interface is the same for all temperatures. As
T increases, however, the amplitude of the oscillations in $F(Y)$ markedly
decreases. This decrease is not merely a reflection of the smaller value of
$\Delta(T)$ in the bulk superconductor. Because of this competition between
thermal and exchange energies, the pair amplitude now has a slightly more
complicated functional form than that given by Eq. (2.22). The amplitude
of the oscillatory decay of $F(Y)$ no longer decreases as the inverse distance
from the interface, but now has an additional slowly varying exponential
term $\Phi'(Y)$, and a purely temperature dependent amplitude, $A(t)$,

$$F(Y) = A(t)\Phi'(Y)\frac{\sin\left[(Y+\theta)/k_{FS}\xi_2\right]}{(Y+\theta)/k_{FS}\xi_2}, \qquad (2.23)$$

where θ is a small, weakly h dependent shift that accounts for the
sharp monotonic decay right at the interface into the ferromagnet. Equation. (2.23) holds for nearly the entire range values of h $0 \le h \le 1$. Certain
exceptions occur in the extreme cases of very small $h \simeq \Delta_0/E_{FM}$ or very
large $h \simeq 1$ and will be addressed below. The form of $A(t)$ in Eq. (2.23),
is fitted well by the form $A(t) = A(0)(1 - t^2)$. Thus $A(t)$ decreases faster
with temperature than the bulk $\Delta(T)$, which shows that the decrease of
the amplitude with temperature is not merely a normalization effect but
involves an intrinsic decrease of the pairing at the interface. Temperature
has a marked effect on the amplitude, but it does *not* wash out the oscillations themselves, which remain quite well defined even at temperatures
quite close to T_c.

The superconductor side (bottom panel of Fig. 2.4) shows a behavior of
$F(Y)$ very similar to that in the textbook [1] $h = 0$ case, with the variation

of $F(Y)$ again occurring over the Ginzburg–Landau length scale $\xi_S(T)$. The very slight wiggles in $F(Y)$ which may be observed near the interface are due to the increased mismatch of the two Fermi energy levels in the ferromagnet with E_{FS}. The effect of the exchange field on $F(Y)$ in the superconductor region therefore seems to be minimal at all temperatures.

In the half metallic case, $h = 1$, where only one spin band is present in the ferromagnet at the Fermi level the characteristic length scale that describes the main oscillatory behavior is still given from Eq. (2.21) as $k_{FS}\xi_2 = 1/\sqrt{2}$, (at $\Lambda = 1$). The relevant spatial variations occur now only on an atomic scale.

The pair amplitude at the interface at constant temperature decreases markedly with h, while at constant h it decreases with T. The minimum value, Δ_{MIN}, of the normalized $\Delta(Y)$ in the superconductor occurs right at the interface, and because of the relatively wide horizontal scale in the bottom panels of Fig. 2.4, it is not possible to read its value from that figure. The effect of the exchange field on Δ_{MIN} turns out to be quite pronounced, and as the temperature approaches T_c, Δ_{MIN} vs t curves tend to collapse into a nearly straight line tending to zero at $t = 1$. The superconducting correlations at the interface are also obviously depleted as h increases.

2.8 Summary

In this chapter I have described in detail the basic procedures that are needed to obtain the exact spectrum of the systems considered in the rest of the book. From the knowledge of the eigenfunctions and eigenvalues one can in principle evaluate any properties of the system. For transport properties we will see in Chaps. 7 and 8 that it is convenient to introduce additional methodologies, but this is for convenience: in principle one can obtain transport properties from the spectrum alone, in any quantum mechanical system.

I have also included in this chapter a discussion of the quantities that determine the proximity lengths for the ordinary singlet pairs.

Chapter 3

Conservation Laws and Triplet Conversion

3.1 Introduction

In this chapter the question of the conversion of singlet pairs to triplets is introduced. As we shall see through the rest of the book, this issue is of the greatest importance and drastically modifies the proximity effects in F/S structures. Its discussion requires first basic considerations concerning conservation laws and other fundamental constraints such as the Pauli principle.

3.2 Conservation Laws

One would intuitively ask whether the rather short proximity lengths such as those discussed in connection with Fig. 2.4 for the *singlet* amplitude would not be much increased were the singlet pairs converted into $S_z = +1$ triplets, lined up with the internal field. For a system including only one F layer the answer is readily found to be negative: the commutator of S_z, the total spin of the Cooper pairs and the Hamiltonian Eq. (2.1) is easily found to vanish. Hence, conversion of singlet pairs to triplet pairs with $S_z = \pm 1$ is, in that case, strictly forbidden by conservation laws. One can show that the quantum number S itself need not be conserved, so that conversion of singlet pairs to triplet pairs with zero z component would be allowed, but such pairs might be subject to the objections discussed in Sec. 3.3.

However, it is quite a different matter if the system includes more than one F layer in such a way that their internal fields are neither parallel nor antiparallel to each other. In that case, the z component of the Cooper pair spin cannot be lined up with both of the internal field axes at the same time, and hence S_z cannot be overall conserved. The simplest case in which it can be achieved is with two F layers only, but in fabricating the sample they must be separated by an N layer, so that one has two well-defined layers. One can then even rotate the magnetization of one layer with respect to the other. This is experimentally the most common case and we will discuss it in Chap. 6. One can also use a single F layer provided that the magnetization structure is non-uniform. Holmium, which has a spiral magnetic structure, has been successfully employed (see Sec. 5.3). It is also possible to use a multiple domain structure in a single layer, but the results are then much harder to interpret and control.

3.3 The Pauli Principle and How to Beat It

Thus, the conservation laws need not stop triplet conversion provided one is able to make more complicated samples or to use more nontrivial materials. But this is not the only fundamental problem involved. One has to look at the spatial dependence also. As long as we are dealing with an ordinary superconductor, spatially s wave Cooper pairs with $S_z = 1$ are forbidden by the Pauli principle.

There is however another way out in this case. The Pauli principle applies to the operator product $\psi_\uparrow \psi_\uparrow$ only when both operators are evaluated at the same time: the anticommutation relations involve a delta function of time. If the average $\langle \psi_\uparrow \psi_\uparrow \rangle$ is odd in time, there is no issue with the Pauli principle. Ordinarily, in defining the Cooper pair amplitudes, one takes the operators at the same time: any time dependence turns out, in fact, to be uninteresting. This is not so for triplet pairs.

The details are a bit intricate and are given in the next section. But at this point we can conclude that odd triplet pairing in a sufficiently complex structure would violate neither conservation laws nor the Pauli principle. This is not all, however: just because something is *not forbidden* it does not mean that it *will happen*. It only will happen if the creation of these pairs would lower the free energy of the system. If not, then no. We will get to that very important point in Chap. 4.

3.4 Odd Triplet Pairing

There is a certain amount of mystification in the literature as to the "oddness" of the pairing. It was originally described in terms of oddness with respect to Matsubara frequencies [7]. It is however much easier and much more intuitive to work in the time domain, rather than in frequency, and find the triplet pairing states that are odd in time.

It is very convenient to define the triplet pairing amplitudes:

$$\mathbf{f_0}(\mathbf{r}, t) = \frac{1}{2}[\langle \psi_\uparrow(\mathbf{r}, t)\psi_\downarrow(\mathbf{r}, 0)\rangle + \langle \psi_\downarrow(\mathbf{r}, t)\psi_\uparrow(\mathbf{r}, 0)\rangle], \qquad (3.1a)$$

$$\mathbf{f_1}(\mathbf{r}, t) = \frac{1}{2}[\langle \psi_\uparrow(\mathbf{r}, t)\psi_\uparrow(\mathbf{r}, 0)\rangle - \langle \psi_\downarrow(\mathbf{r}, t)\psi_\downarrow(\mathbf{r}, 0)\rangle]. \qquad (3.1b)$$

We will repeatedly verify below that these amplitudes vanish at $t = 0$, as required by the Pauli principle. In the quasi 1-d geometry, the spatial dependence on $\mathbf{r}$ becomes simply a dependence on y. $\mathbf{f_0}$ corresponds to $S_z = 0$ and $\mathbf{f_1}$ to the more interesting case of $S_z = \pm 1$.

To sort out the time dependence of operator quantities it is most convenient to use the Heisenberg picture. This is something that many physicists learn in their first Quantum Mechanics course and then, for unfathomable reasons, forget. Thus we write ψ_ς in the Heisenberg representation:

$$\psi_\sigma(t) = e^{(i\mathcal{H}_{\text{eff}}t)}\psi_\sigma e^{(-i\mathcal{H}_{\text{eff}}t)}. \qquad (3.2)$$

To put this in terms of the quasiparticle amplitudes, we apply Eqs. (2.5) and the transformation Eqs. (2.6). We recognize the γ's as the time-dependent operators and u_n and v_n as c-numbers, allowing us to write e.g., (dropping the spin subscripts for simplicity) $\psi(t) = \sum_n (u_n\gamma_n(t) \mp v_n\gamma_n^\dagger(t))$. We can then immediately write down the Heisenberg equations of motion for the γ's as

$$i\frac{\partial \gamma_n}{\partial t} = [\gamma_n, \mathcal{H}_{\text{eff}}] \qquad (3.3)$$

and

$$i\frac{\partial \gamma_n^\dagger}{\partial t} = [\gamma_n^\dagger, \mathcal{H}_{\text{eff}}]. \qquad (3.4)$$

These equations of motion, given Eqs. (2.6), have trivially the solutions:

$$\gamma_n(t) = \gamma_n e^{-i\epsilon_n t} \qquad (3.5)$$

and

$$\gamma_n^\dagger(t) = \gamma_n^\dagger e^{i\epsilon_n t}. \qquad (3.6)$$

By substituting these results into the above equations for $\mathbf{f_0}$ and $\mathbf{f_1}$, taking into account Eqs. (2.5) and the fermionic character of the Bogoliubov operators, one obtains immediately:

$$\mathbf{f_0}(y, t) = \frac{1}{2} \sum_n [u_{n\uparrow}(y) v_{n\downarrow}(y) - u_{n\downarrow}(y) v_{n\uparrow}(y)] \zeta_n(t), \tag{3.7a}$$

$$\mathbf{f_1}(y, t) = -\frac{1}{2} \sum_n [u_{n\uparrow}(y) v_{n\uparrow}(y) + u_{n\downarrow}(y) v_{n\downarrow}(y)] \zeta_n(t), \tag{3.7b}$$

where $\zeta_n(t) \equiv \cos(\epsilon_n t) - i \sin(\epsilon_n t) \tanh(\epsilon_n / 2T)$.

The above procedure is indeed simpler, more elementary, and more elegant than dealing with the Matsubara frequencies.

The expressions Eq. (3.7) are in general complex. The amplitudes depend on time (and vanish at $t = 0$ of course). It is convenient and natural to measure the time in terms of the cutoff (Debye) frequency, and we define $\tau \equiv \omega_D t$, as seen in Table 2.1.

3.5   Existence of Triplet Amplitudes

In the above, we have shown that the existence of proper (odd) triplet amplitudes is *possible* and we have developed a formalism for describing them. Now, we all know that just because something is *allowed* it does not mean that it *must* happen. It bears repeating that it will happen if and only if converting singlet to triplet pairs lowers the free energy. While the question of the thermodynamics and the free energy will be discussed in Chap. 4 the basic result is very simple and can be described here. The solutions of the basic (Bogoliubov-DeGennes) equations are known[2] to correspond to local minima of the free energy. So, in principle one simply has to follow the self-consistent procedure described in Chap. 2 and obtain the eigenfunctions and eigenvalues. One then substitutes the results in Eq. (3.7) and simply checks if the result is zero or nonzero. In all relevant cases,where the conservation laws do not forbid triplets, it turns out to be nonzero, the only exception being if the barriers are so high that one is in the extreme tunneling limit. In that case, however, one has no proximity effects at all.

A separate question is the *size* of the triplet correlations. One has to be careful about characterizing the size since the triplet correlations are complex amplitudes and, furthermore, depend on position and on time. The position dependence is of course also a function of geometry and everything depends, in the end, on all parameters presented in Table 2.1. One can however make the general statement here, which will be corroborated in some of the examples discussed in this and subsequent chapters, that a percentage of order 5 to 10% of the original singlet pairs typically gets transformed into triplet pairs, with roughly half of them being of the $S_z = \pm 1$ variety, relevant to the longer range proximity effect.

Below, I discuss one specific example. As we will see, the existence of triplet pairs have far-reaching consequences and additional examples will appear later in the book.

3.6 Some Triplet Amplitude Results: An Example

For the pedagogical reasons discussed above, here I include as an illustration some simple results for the triplet amplitudes. More detailed results in less elementary and experimentally more relevant situations will be presented in Chap. 5. In this simple example (taken from Ref. [8], with the corresponding figures) there are three layers: F_1, S, and F_2. The general shape of the structure is shown in Fig. 3.1. For this specific example, the S layer has $D_S = 200$ (which is twice the coherence length chosen) and $D_{F_1} = D_{F_2} = 250$. The most important parameters are the angle α between the F_1 and F_2 magnetizations and the magnet dimensionless strength h. In this example there are no scattering barriers included, so that proximity effects are maximized. The internal fields in F_1 and F_2 are tilted by angles $\pm\alpha/2$ with respect to the z-axis of quantization. If we take $\alpha = 0$ or $\alpha = \pi$, then the magnetizations of both layers lie along the same direction, parallel or antiparallel. We can find the quasiparticle amplitudes

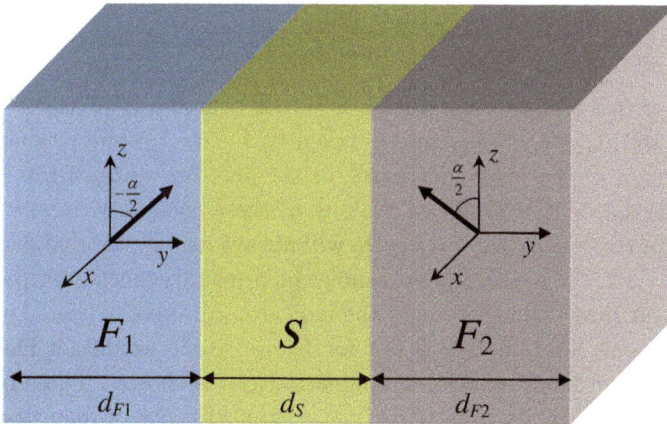

Fig. 3.1 Schematic of the FSF trilayer discussed. The left ferromagnetic layer denoted as F_1 has a magnetization oriented at an angle $-\alpha/2$ in the $x-z$ plane, while the other magnet F_2, has a magnetization orientation at an angle $\alpha/2$ in the $x-z$ plane.

on a different quantization axis in the $x - z$ plane forming an angle α' with z, by performing a spin rotation via the matrix:

$$\widehat{U}(\alpha') = \cos(\alpha'/2)\hat{1} \otimes \hat{1} - i\sin(\alpha'/2)\rho_z \otimes \sigma_y, \qquad (3.8)$$

where we have again used ρ as a set of Pauli-like matrices in particle-hole space. From the arguments in Secs. 3.2 and 3.3 it follows that the induced amplitude $\mathbf{f_1}(y, t)$ may exist (at finite times) only at nonzero α, while $\mathbf{f_0}(y, t)$ is allowed at any α. For $\alpha = \pi$, when the magnetizations are antiparallel and along the x-axis, no triplet amplitudes with nonzero component along that axis can exist. The matrix $\widehat{U}(\alpha)$ in Eq. (3.8) can be used to verify this by performing the corresponding spin rotations.

To carry out the calculations, one must perform the the diagonalization of the Hamiltonian as explained in Secs. 2.5.1 and 2.5.2. It is important, as we shall see below, that the diagonalizations be iterated until self-consistency is achieved. Then, by substituting the resulting wavefunctions in Eq. (3.7) one obtains the triplet amplitudes. One can then extract the triplet (as well as the singlet) amplitudes, and examine them. In anticipation of things to come, the necessary matrix elements for this problem were already given as an example: see Eq. (2.19).

Some results for the amplitudes are given below. Since they are somewhat intricate, it is convenient and useful to describe the range of triplet penetration by introducing characteristic proximity lengths ℓ_i as,

$$\ell_i = \frac{\int dy |f_i(y, t)|}{\max |f_i(y, t)|}, \qquad (3.9)$$

where $i = 0, 1$ and the integration is either over the S or the F region. In Eq. (3.9), $|f_0|$ and $|f_1|$ can be taken to be the absolute value of the real parts, or of the imaginary parts, or the absolute value of the complex quantities $|\mathbf{f_0}|$ or $|\mathbf{f_1}|$. One can use any of these choices with very similar results. In the results shown $f_i(y, t)$ will always denote the real part.

In Fig. 3.2, the top two panels show the proximity lengths within the F material, at three values of h and the indicated values of α. For the $S_z = 0$ amplitude one can take $\alpha = 0$ but for $S_z = \pm 1$ one would get identically zero and the case of orthogonal magnetizations is chosen. These opposite magnetizations are along the x-axis, *not* along the axis of quantization, z. The results are plotted as a function of τ. The amplitudes themselves vanish at $\tau = 0$ (see Fig. 3.4) although this is somewhat obscured in Fig. 3.2 because in that limit both the numerator and the denominator in the right

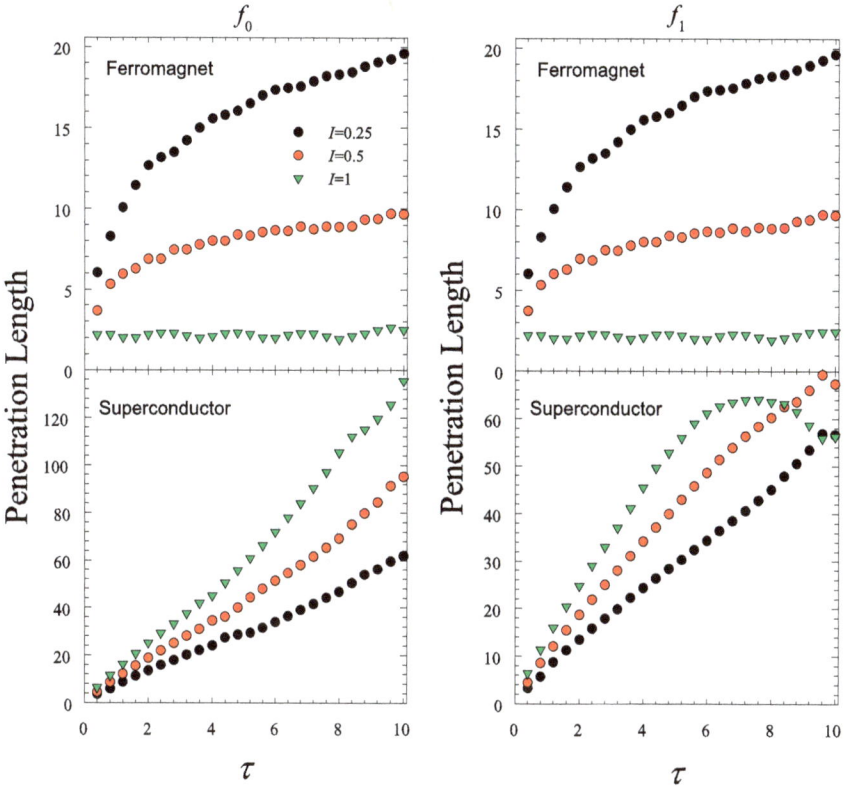

Fig. 3.2 Proximity depths for the triplet amplitudes (see Eq. (3.9)), plotted as a function of dimensionless time τ, as calculated from f_0 and f_1 (the real parts of $\mathbf{f_0}$ and $\mathbf{f_1}$) in both the S and F regions, for the values of h indicated. The magnetizations are aligned in the left panels and form an angle of $\pi/2$ in the right panels.

side of Eq. (3.9) vanish. The results for this quantity are not too dissimilar whether they are calculated from f_0 or from f_1. The proximity length at constant time decreases with h and it shows signs of saturating with time at a value which for $h = 0.25$ approaches that of the superconducting coherence length. On the S side the situation is very different: the results for f_0 and f_1 are clearly dissimilar with the penetration length for the former quantity being the larger one. This result, at first surprising, arises from the geometry and from the magnetization projections of each F layer being in opposite $\pm x$ directions, forcing the triplet f_1 to possess a node at the center of the trilayer. No such requirement exists for f_0, as it is spatially symmetric. Except for the case of f_1 at $h = 1$, there are no signs of

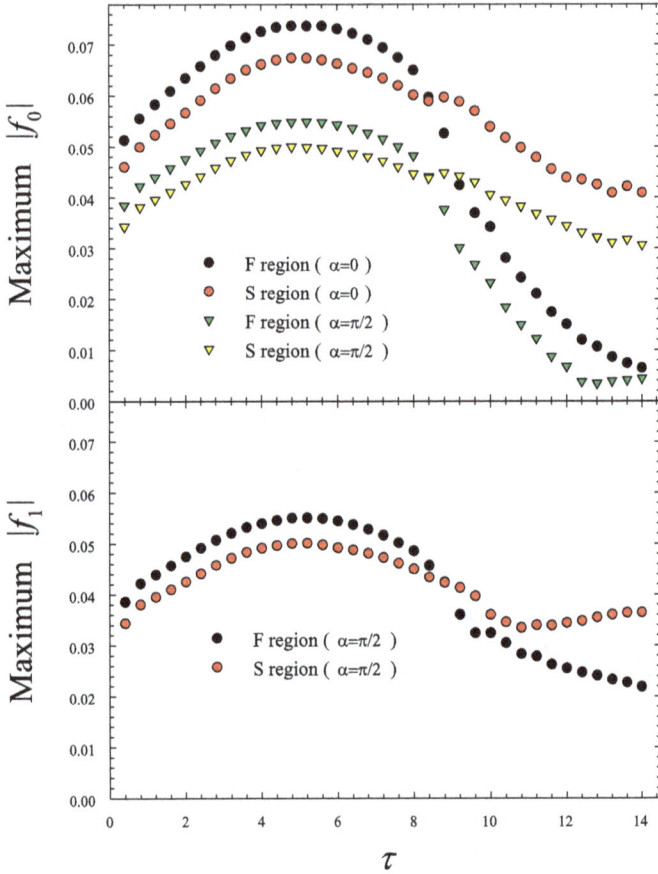

Fig. 3.3 Maximum absolute values of f_0 and f_1 (see text) as a function of dimensionless time τ at $h = 1$. In the top panel we consider f_0 for both the aligned and anti-aligned cases while in the bottom panel we consider f_1 at a misalignment angle of $\pi/2$.

saturation in the τ range included. The maximum value of τ displayed corresponds to the case in which the entire intrinsically singlet superconductor layer, two coherence lengths thick and sandwiched between two magnets, is nearly entirely pervaded (see Fig. 3.4) by induced triplet correlations.

The evolution of the spatial maximum values of $|f_0(y, t)|$ and $|f_1(y, t)|$ with τ is shown in Fig. 3.3. This maximum value is typically attained near the interface, in either the F or S regions. In this figure, the magnets are half metallic, $h = 1$. In the top panel, the results for $|f_0|$ are shown at both $\alpha = 0$ and $\alpha = \pi/2$. At earlier times, the value $|f_0(y, t)|$ at its peak, which is just inside the F region, exceeds the maximum value of this quantity in

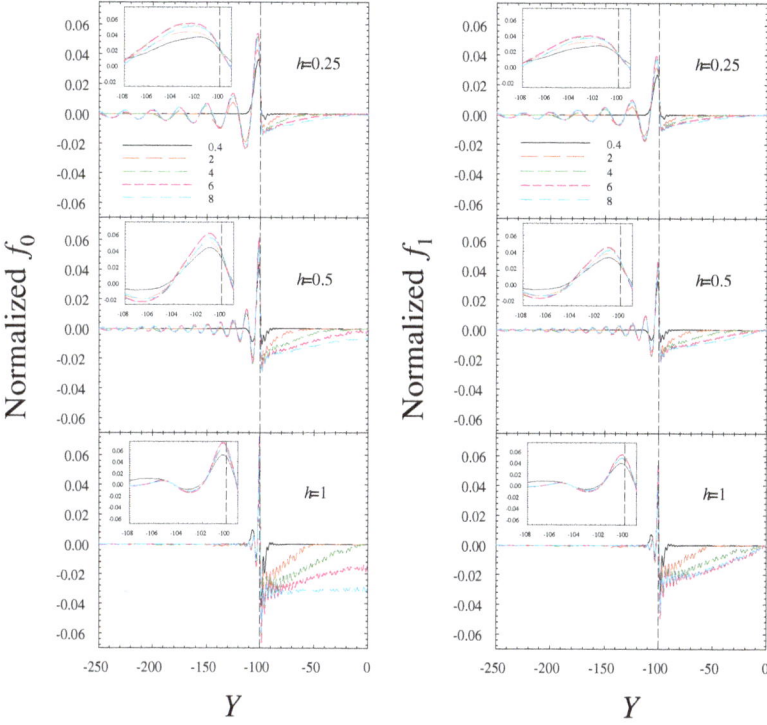

Fig. 3.4 The real parts, f_0 and f_1, of the triplet pair amplitudes $\mathbf{f_0}$ and $\mathbf{f_1}$ (Eq. (3.7)), plotted vs. Y for three values of h at different times τ indicated in the legends of the top panels. These quantities are normalized to the singlet amplitude in bulk S material. The main plots show (left side of the vertical dashed line) half of the S region and part of the F_1 region. The insets are blow-ups of the region near the interface. The misalignment angle is zero in the left panel and to $\pi/2$ in the right panel.

S (see below, in Fig. 3.4). At longer times, however, there is a crossover as the size of the peaks in F decreases rather sharply, as explained above, while the size of the amplitude in S decreases only slowly, as the triplet correlations fill the S layer. It is apparent from careful examination of the $h < 1$ panels in Fig. 3.4, that this crossover does not occur for smaller values of h in the τ scale included here. At long times the triplet correlations essentially fill the S layer. In the bottom panel, results for $|f_1|$, are shown, of course only at $\alpha = \pi/2$ since this quantity vanishes identically for collinear magnetizations. The results are clearly very similar except that the results in the S region appear to saturate and do not decrease at long times. In all cases the maximum value of the quantity plotted crests near $\tau = 6$.

It is of course most important to see the behavior of the amplitudes themselves, as a function of position. Since the triplet amplitudes are (see Eq. (3.7)) complex quantities one must in principle examine their real and imaginary parts separately. However, their behavior is similar and we will look here at the real parts only. In Fig. 3.4 we can see comprehensive plots of the real parts of $\mathbf{f_0}(y, t)$ and $\mathbf{f_1}(y, t)$, which are here denoted, we recall, as $f_0(y, t)$ and $f_1(y, t)$ respectively. The behavior of the imaginary parts is actually quite similar. These quantities are plotted in terms of the dimensionless variable $Y \equiv k_F y$. The amplitudes are normalized to the value of the usual singlet amplitude in a pure bulk S sample. The temperature is set to zero in this figure. In the main plots, half of the S region and a portion (three fifths) of the left (F_1) region are included. The corresponding portion on the F_2 side can be inferred from the geometry and symmetry considerations. Results are plotted at three values of h and at a number of finite times τ between 0.4 and 8 as indicated in the legends. At $\tau = 0$ the computed triplet amplitudes vanish identically, in agreement with the Pauli principle. This is true, however, *only when the calculation is performed to self-consistency: non-self-consistent results invariably violate the Pauli principle* near the interface. The results for f_0 are given at an angle $\phi = 0$ while those for f_1 are at $\phi = \pi/2$. At $\phi = 0$, f_1 vanishes identically since the z component of the total spin is then a good quantum number. At $\phi = \pi/2$, and short τ scales, the spatial dependences of the two triplet components coincide, albeit with different signs in the two magnet regions, due to the magnetization vectors having equal projections on the x and z-axes. At longer times, when the triplet amplitudes extend throughout the S layer and couple the two magnets, f_0 and f_1 deviate from one another. The insets in each panel amplify and clarify the region near the interfaces.

On the F side, both amplitudes peak very near the interface and then decay in an oscillatory manner, reminiscent of the behavior (as seen in Chap. 2) of the usual pair amplitude. Although the height of the first peak does not depend strongly on h, the subsequent decay in the F material is faster for larger values of h. This can be attributed to a decreased overall proximity effect: here it is assumed that at $h = 0$ there would be no mismatch between the Fermi surface wavevectors of the two materials, implying that as h increases the mismatch between both the up or the down Fermi wavevectors $k_\uparrow$ and $k_\downarrow$, on the F side, and that in the S side increases. The location of this first peak depends very clearly on h, its distance to the interface decreasing as h increases as approximately $1/h$ consistent with the general rule that the oscillatory spatial dependences on the F side are determined by the inverse of $k_\uparrow - k_\downarrow$. The height of the first peak depends strongly on time and is maximum at times τ of about 2π. It is quite obvious that at intermediate values of h the penetration of the triplet correlation into the F material is rather long ranged.

On the superconducting side the behavior is quite different: the triplet correlations penetrate into the S material over a distance that rather quickly reaches two correlation lengths and then of course saturates at the sample size, without signs of decaying in time at these length scales. Furthermore this effect now increases sharply with h and is maximal in the half metallic case. Thus, the magnets *act as sources of triplet correlations* that enter the S material and this effect is stronger when h is larger.

I conclude this example by showing temperature dependence of the triplet amplitudes, focusing again on the real part. This is done in Fig. 3.5. There the peak values of $|f_0|$ and $|f_1|$ (at $\tau = 4$) are shown as a function of T. These quantities, which refer to the absolute value of the real parts, not to the absolute value of the complex quantities, are determined by calculating $\max\{f_0(y, T)\}$ and $\max\{f_1(y, T)\}$ throughout the structure for a given temperature. The inset depicts the corresponding peak values of the ordinary self-consistent equal-time singlet pair amplitude. This singlet

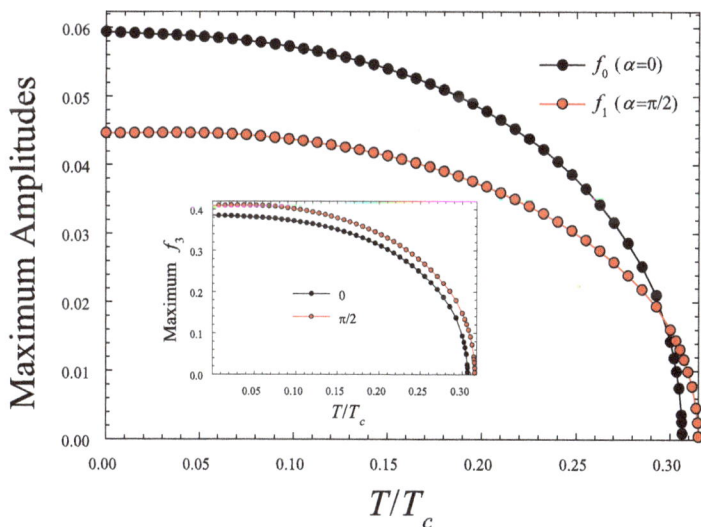

Fig. 3.5   Maximum values of the real parts of the triplet amplitudes at $h = 0.5$ and $\tau = 4$, as a function of temperature and indicated values of α. Inset: the corresponding peak values of the ordinary singlet amplitude, $f_3 \equiv \Delta(y)/g$.

amplitude is, it should be noted, substantially smaller than unity: sandwiching a superconductor between magnets does the superconductivity no good. For both the triplet and singlet amplitudes, there is a strong dependence on T as the temperature approaches the system's transition temperature, which is about $0.32T_c$ for this example. Technically, the determination of the self-consistent amplitudes is more difficult at higher T, when the number of iterations is in principle much higher. This increase in computational time can be reduced by up to an order of magnitude by taking as the initial spatial pair potential at a given T the result for the previously obtained next lower temperature times a T dependent factor derived from the linearized Ginzburg–Landau theory.

In Chapter 4, we will consider more generally, the calculation of T_c and the thermodynamics of these systems.

Chapter 4

Thermodynamics

4.1 Basic Ideas

The equilibrium properties of the systems that we are considering depend very drastically not only on the temperature, but also on the geometry. Imagine for example a trilayer system consisting of two identical outer S layers separated by a thin F layer. The oscillations of the singlet pair amplitude extending (see Fig. 2.1) from each of the S layers into the F layer, from the left and from the right, will have to adjust so as to somehow match in the middle. It is intuitively obvious, and in fact correct, that this adjustment will depend on the thickness d_F. This means that the thermodynamic properties of such a system will depend on d_F since they must depend on the pair potential itself.

But there is more to it: consider the just mentioned trilayer consisting of two superconductors separated by an intervening magnet, in a case involving very high interfacial barriers, so that there is no proximity effect: the two superconductors do not really know about each other. If they are thick enough and the temperature is low, then the order parameter will be approximately a constant, in each one of them. The absolute value of the constant will be the same, but how about the phase? Assuming, as we do, that no current flows, then the relative phase could be 0 or π. In other words, the relative sign could be the same or the opposite. These two states are in this limit obviously degenerate. The pair potential is zero in the intermediate layer.

Now imagine that the interfacial scattering barriers are lowered, so that proximity effects start to be felt. Then the pair potential no longer vanishes in the intermediate layer, and the states with the zero and π symmetry are in general no longer degenerate.

There of course will be a limit where the proximity effect is strong and only one of the two states will correspond to a free energy minimum. But there may be more than one local minimum. To see this consider the iterative procedure described in general in Sec. 2.5.1, which stops when a local free energy minimum is reached. If there is more than one local free energy minimum these local minima can be reached by starting with a different initial guess. For the high barrier trilayers we just mentioned as an example here, if one starts with a zero-like guess one reaches the zero state. If one changes the sign in one of the two S layers, then the π state is reached. If there is only one minimum, that minimum is always reached although with the "wrong" initial guess it takes many more iterations to do so. It is worth spending the time and money to find this out.

How does one find the global minimum when there is more than one local free energy minimum? There is only one way: after determining the local minima one must find out which one corresponds to the lowest free energy. This is very easily said, and not so easily done, and we will see in this chapter that it can be done and how we can do it.

In addition to the geometry, one must take into account the possibility that the temperature takes also a role: the nature of the global minimum may change with T. In that case there may be a first order transition occurring as the temperature changes. We will consider this too in this chapter.

4.2 The Free Energy

So, one has to bite the bullet and calculate the free energy or, to be more precise, the condensation free energy, defined as the difference between the free energies in the normal and the superconducting state. For bulk superconductors, the expression for the condensation free energy can be found in textbooks. For an inhomogeneous superconductor, one also finds expressions in several books and papers. These expressions often appear to differ from each other but one can painfully prove, by imaginative use of self-consistency conditions, that they are all (well, nearly all) correct. For computational convenience I know nothing that beats the expression in Ref. [9]:

$$\mathcal{F}(T) = -2T \sum_{n}{}' \ln\left[2\cosh\left(\frac{\epsilon_n}{2T}\right)\right] + \frac{1}{d}\int_0^d \frac{\Delta^2(y)}{g(y)}dy, \qquad (4.1)$$

where d is the total thickness of the system in the y-direction and the primed sum is as defined in connection with Eq. (2.12). This expression is derived in a few steps by starting out with the diagonalized Hamiltonian, which is of the free-fermion form, and the standard expression for the free energy of a free Fermi gas. A big technical advantage of the expression Eq. (4.1), is that only the energy eigenvalues appear explicitly; the eigenfunctions are involved only indirectly through the self-consistent $\Delta(y)$. This makes it computationally convenient. It is amusing to note that the reference mentioned goes on to derive other expressions which are allegedly more computationally convenient but that, in reality, are not so for our purposes. It is equivalent to several other expressions found in the literature which contain the quasi-particle amplitudes explicitly.

The condensation free energy is $\Delta\mathcal{F}(T) \equiv \mathcal{F}_S - \mathcal{F}_N$, where $\mathcal{F}_S$ is the free energy of the superconducting state and $\mathcal{F}_N$ that of the non-superconducting system. One computes $\mathcal{F}_N$ by setting $\Delta \equiv 0$ in the basic equations (2.8) and (4.1). Calculating $\Delta\mathcal{F}(T)$ is a significant numerical challenge: recall that in a bulk superconductor [3] at $T = 0$ one has $\Delta\mathcal{F}(0) = -(1/2)N(0)\Delta_0^2$, where $N(0)$ is the usual density of states and Δ_0 is the order parameter for the bulk superconductor at $T = 0$. We see that $\Delta\mathcal{F}(T)$ is several orders of magnitude smaller than $\mathcal{F}_N \propto N(0)\omega_D^2$. Hence, to obtain $\Delta\mathcal{F}$ one must subtract two numerically obtained large quantities in order to extract a difference several orders of magnitude smaller than the terms subtracted. Furthermore, to make things even more interesting, the difference in condensation free energies of competing self-consistent states (when they occur) is a fraction of the condensation free energy of each of them. To obtain sufficiently accurate values of $\Delta\mathcal{F}$ requires therefore a very high degree of precision in calculating $\mathcal{F}_S$ and $\mathcal{F}_N$, so that we can distinguish the relatively small differences between competing states to locate phase transitions. This implies high precision in both $\Delta(y)$ and the energy spectrum. This situation is made more challenging by the need to calculate derivatives of $\Delta\mathcal{F}$ to obtain thermodynamic functions and latent heats. It is however all quite doable, as we shall see presently, but one must be careful.

It is convenient to normalize the condensation free energy to twice its absolute value in bulk S material at $T = 0$ which, as mentioned above, is $N(0)\Delta_0^2$. Once one knows the free energy, all other thermodynamic functions can be found. For example the dimensionless condensation entropy $S(T)$, can be defined as the negative derivative of the normalized $\Delta\mathcal{F}(T)$ with respect to the reduced temperature $t \equiv T/T_c^0$ (see Table 2.1):

$$\Delta S \equiv -d\Delta\mathcal{F}/d(T/T_c^0), \tag{4.2}$$

and the dimensionless condensation energy $U(T)$ as

$$U(T) \equiv \Delta\mathcal{F}(T) + (T/T_c^0)S(T). \qquad (4.3)$$

These definitions will be used below.

4.2.1 *The calculation of the transition temperature*

As a preliminary step, I discuss here the calculation of the transition temperature T_c in F/S heterostructures. This is always smaller than the temperature at which bulk S alone becomes superconducting, T_c^0. The transition temperature T_c from the non-superconducting to the superconducting state can be numerically calculated by evaluating $\Delta\mathcal{F}$ as a function of T on a fine grid and finding the temperature at which it vanishes. However, it is much easier to evaluate T_c directly by treating $\Delta(y)$ as a small parameter and linearizing the self-consistency equations. In this way the calculation is nearly entirely analytic. This is the way the transition temperature in the bulk is calculated in textbooks.

To do so the amplitudes are written as $u_n^\uparrow(y) = u_n^0(y) + u_n'(y)$ and $v_n^\downarrow(y) = v_n^0(y) + v_n'(y)$ (dropping some spin indices for notational simplification). The $u_n^0(y)$ and $v_n^0(y)$ terms are computed from the zeroth order equation, which is obtained by setting $\Delta \equiv 0$ in Eq. (2.8). The form of the zeroth order equation implies that $u_n^0(y)$ and $v_n^0(y)$ are completely decoupled and have distinct energy spectra, denoted by ϵ_n^p and ϵ_n^h respectively: particles and holes are decoupled in the normal state. Proceeding to calculate the lowest order corrections, one incorporates quasiparticle coupling through the pair potential matrix. One can then obtain $u_n'(y)$ and $v_n'(y)$ from textbook quantum mechanical perturbation formulas. The intermediate sums are in principle over the entire zeroth order spectrum, but as a practical matter it is enough to include in these sums energies ϵ_m^p and ϵ_m^h within a few ω_D of the Fermi level.

One can then expand the quasiparticle amplitudes and their first order corrections in a sine wave basis $\phi_q(y) = \sqrt{2/d}\,\sin(k_q y)$, with $k_q = q\pi/d$, as done before (see Eqs. 2.14 and 2.15), i.e., $u_n^0(z) = \sum_q^N u_{qn}\phi_q(z)$ $v_n^0(z) = \sum_q^N v_{qn}\phi_q(z)$, $u_n'(z) = \sum_q^N u_{qn}'\phi_q(z)$, and $v_n'(y) = \sum_q^N v_{qn}'\phi_q(y)$. The range of the sums over k_q is formally infinite, but as in Sec. 2.5.1 it is only necessary to sum up to a wavenumber k_N with an associated energy a few ω_D from E_F. Inserting these expansions into Eq. (2.12) gives the lowest order correction to $\Delta(y)$, which one then expands in the sine function basis. Upon taking into account the orthogonality of the basis functions, the expanded Eq. (2.12) is then transformed into a matrix equation

$$\Delta_l = \sum_k J_{lk}\Delta_k, \qquad (4.4)$$

where the Δ_k are the expansion coefficients of $\Delta(y)$ in terms of $\phi_q(y)$. One finds after straightforward algebra the following expression for J_{lk}:

$$
\begin{aligned}
J_{lk} = \frac{gN(0)}{4\pi} \\
\times \int d\varepsilon_\perp {\sum_n}' \Bigg\{ \sum_m \sum_{pq}^N u_{pn}^0 v_{qm}^0 K_{pql} \frac{\sum_{ij}^N v_{im}^0 u_{jn}^0 K_{ijk}}{\epsilon_n^p - \epsilon_m^h} \tanh\left(\frac{\epsilon_n^p}{2T}\right) \\
+ \sum_m \sum_{pq}^N v_{pn}^0 u_{qm}^0 K_{pql} \frac{\sum_{ij}^N u_{im}^0 v_{jn}^0 K_{ijk}}{\epsilon_n^h - \epsilon_m^p} \tanh\left(\frac{\epsilon_n^h}{2T}\right) \Bigg\}. \qquad (4.5)
\end{aligned}
$$

Here I have used $gK_{ijk} = \int_0^d g(y)\phi_i(y)\phi_j(y)\phi_k(y)dy$. The sum over transverse energies is transformed in this analytic calculation into an integral over $\varepsilon_\perp$ which reflects the dependence of the zeroth order quasiparticle amplitudes and energies on $\varepsilon_\perp$. The sum over n is here over longitudinal quantum numbers (see remarks below Eq. (2.12)). The prime denotes the usual limitation indicated below Eq. (2.12) on the energies, ϵ_n^p and ϵ_n^p.

The transition temperature can then be found by treating Eq. (4.4) as an eigenvalue equation for the matrix J_{lk}. At the transition temperature T_c the largest eigenvalue is unity, while if $T > T_c$ all eigenvalues are less than unity. One inputs some guess on the higher side for T_c/T_{c0} and then lowers T until the largest eigenvalue becomes larger than unity. Alternatively, one begins with a low T guess and increases it until the largest eigenvalue becomes unity. Algorithms to find only the largest eigenvalue of the matrix exist and are very efficient: you should not think of finding all eigenvalues and then checking for the largest! This procedure does not require an iterative process and only the last step (finding the eigenvalue) must in practice be performed numerically. Therefore this method is much more efficient than checking for the vanishing of the condensation free energy, and it also provides a check on the numerics of the free energy itself. A comparison of the results of this calculation with experimental data will be presented in section Sec. 4.3.3 below.

4.3   Example: SFS Trilayers

The above is much better understood in the context of an example, and for this purpose we will consider an SFS trilayer, as discussed above: see Sec. 4.1. The figures in this example are taken from Ref. [10]. I will first discuss the results for the thermodynamics and the phase transitions that ensue. This will include a detailed discussion of the phase diagram for this trilayer in the most interesting region of the three dimensional space spanned by T, the mismatch parameter Λ and the F layer thickness. A discussion of the properties of the transition temperature T_c as a function of d_F and a comparison with experiment follow. For completeness, other quantities of interest, including the magnetization (reverse proximity effect) will also be discussed. In this example, the S layer thickness is fixed to $D_S \equiv k_F s d_S = 100$, and the BCS coherence length ξ_0 is equal to d_S. For d_S of order of or larger than ξ_0, results are only weakly dependent on d_S. Hence, the results are applicable to a very wide range of values of this variable, provided d_S is not too small. The dimensionless thickness $D_F \equiv k_F s d_F$ of the ferromagnetic layers is varied over the range of interest, which corresponds to relatively small values, since at large ones the F/S proximity effects are less important. The magnetic strength parameter is taken to be $h = 0.2$ unless otherwise noted. The effects of varying h are physically similar to those of varying D_F since the pair amplitude oscillations in F are governed (as we have seen in Chap. 2) by the difference $(k_\uparrow - k_\downarrow)d_F$ between Fermi wavevectors in the spin bands in F. The Fermi wavevector mismatch parameter, Λ, to which results are quite sensitive, will be varied over the experimentally relevant range $0.1 \leq \Lambda \leq 1$. Table 2.1 should be consulted for all normalizations and definitions.

4.3.1   *Thermodynamic functions*

I will start with the thermodynamics and see how it leads to the system's phase diagram. This requires the calculation of the dimensionless condensation free energy. The first question is whether there are transitions between the 0 and π states as the temperature varies. In Fig. 4.1 the free energy is shown for two cases where such a transition occurs. The parameters are indicated in the caption. For these parameter values it turns out that both the 0 and the π states are *local* free energy minima and hence only the quantitative evaluation of the free energy can tell us which is the stable state. We see that a transition exists from the π state at low T to a higher temperature 0 state, the transition occurring at $t = 0.15$ in the top panel and at $t = 0.23$ in the bottom one.

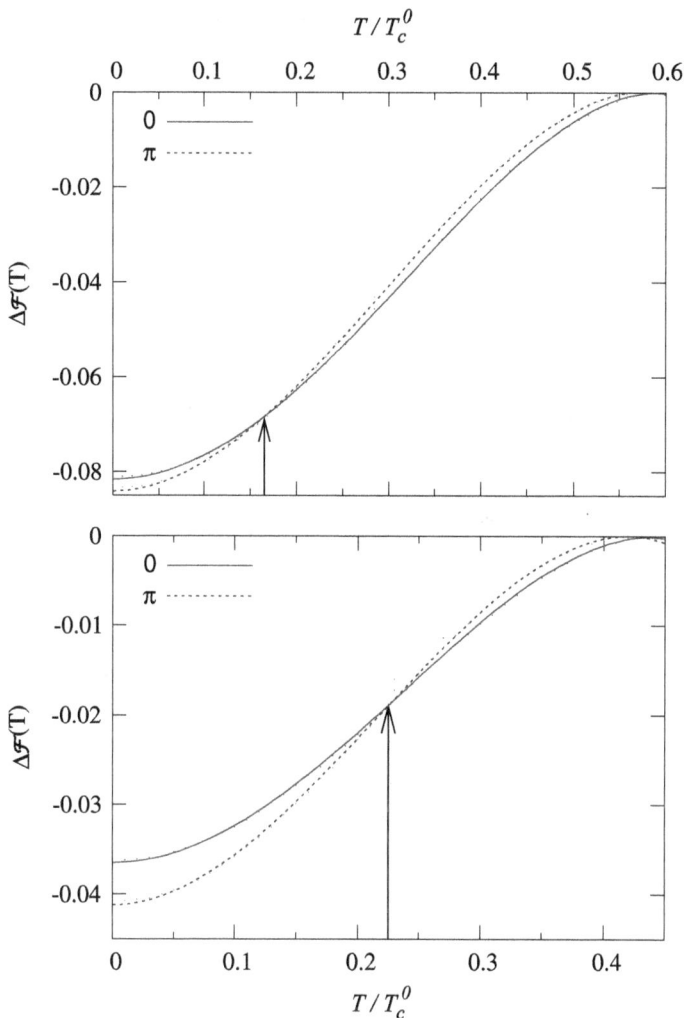

Fig. 4.1 The normalized (see text) condensation free energy as a function of temperature for an SFS trilayer. In both cases both the zero and π states are local free energy minima and a π to 0 transition occurs upon increasing T (arrows). In the top panel $\Lambda = 0.550$ and $D_F = 7.1$ while in the bottom panel $\Lambda = 0.65$ and $D_F = 5.9$.

Examining such free energy curves, we see that their slope is zero at both $T = 0$ and at T_c. The first reflects the third law of thermodynamics, while the second tells us that the transition between the superconducting and the normal states is of second order. This transition occurs at about $t = 0.45$, this being of course lower than $t = 1$, the transition temperature

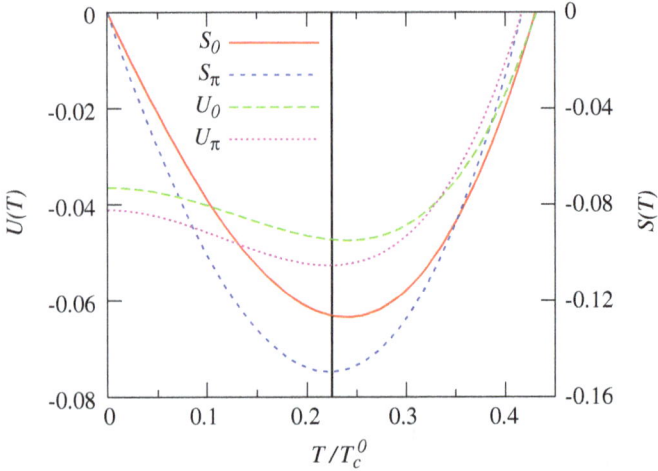

Fig. 4.2 The normalized thermodynamic functions, condensation energy and entropy, (see text) for the system in the bottom panel of the previous figure. Results are given for both the 0 and π states. The transition temperature between them is marked by the heavy line.

of bulk S material. As to the π to 0 transition it is on the other hand of first order. You can see that because there is an obvious discontinuity in the derivative of the free energy. This discontinuity represents a latent heat.

From the condensation free energy results, one can then easily proceed and evaluate numerically the other thermodynamic functions, using elementary formulas. The normalized (see above) condensation entropy and energy for the case presented in the bottom panel of Fig. 4.1 are plotted in Fig. 4.2. The free energy crossing at the first order transition is seen to arise from energy and entropy balance: one cannot call the transition either "energy driven" or "entropy dominated". It is a combination.

We can also see how easy it is to extract from such results the latent heat, which is simply the difference in the entropy values as the transition is crossed. Indeed, the signature of a first order phase transition is its latent heat. It is physically very sensible to normalize this latent heat by the value of the specific heat of a normal bulk sample of S material at $T = T_c^0$. This is because the latter quantity is, within the free electron model, the same as the entropy. This normalized quantity is of the order of one percent, which means that this latent heat is measurable, although nobody seems to have tried.

As an example, in Fig. 4.2 we show results for the dimensionless latent heat L defined and normalized as explained above. Results are plotted as

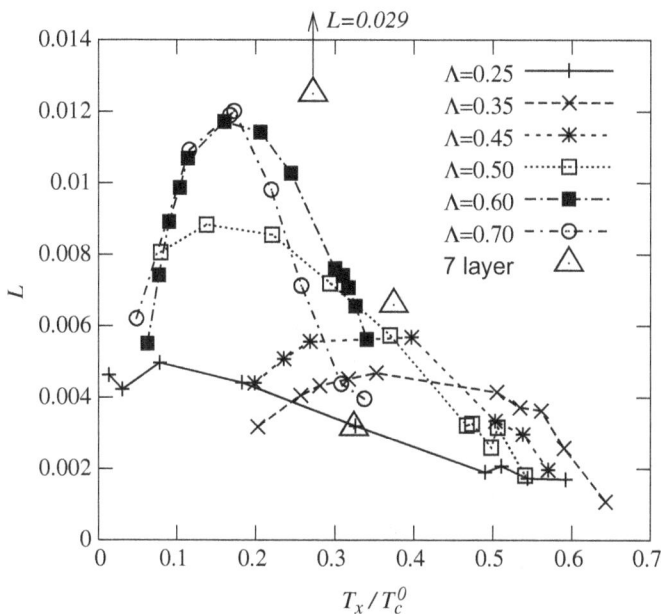

Fig. 4.3 Latent heats L, of the first order $0/pi$ transition, normalized as explained in the text. It is plotted against the reduced temperature of the first order phase transition, the crossing temperature T_x. The symbols joined by lines are the relevant ones, for an SFS trilayer: the value of T_x is changed along the horizontal axis by varying D_F, and from curve to curve by varying Λ (see legend). The two triangles are for a seven layer system not discussed in the text. The vertical arrow attached to the topmost triangle indicates that it corresponds to a value $L = 0.029$ (off the scale).

a function of T_x, that is, the crossing temperature. Nearly all the results shown (see caption) are for a single junction: to vary the transition temperature T_x one changes D_F. Results are included for several different values of Λ, as indicated by the symbols connected by straight segments. The two data points indicated by the isolated triangles correspond to transitions not discussed in this book (see Ref. [10]) in a seven layer $SFSFSFS$ structure. One of them corresponds to a value larger, by over a factor of two, than the upper end of the scale.

The latent heats vanish as T_x approaches 0 or T_c, this is of course consistent with the smaller condensation entropy of each state in those limits. However, whenever T_x does not approach these limits the latent heat can exceed 1% of $C_n(T_c^0)$ for one junction, and even more for the three junction system. Since L is an extensive property, it should be easier to observe these latent heats in larger systems. A value of $L \approx 0.01$ would correspond to picojoules in actual samples of relatively small size [11]. Such latent heats can be readily observed via standard techniques used to measure specific and latent heats in films. Even smaller specific heats can be measured using multiple samples: attojoule level results have been reported [12] in electronic systems. We see therefore that whenever a first order transition occurs, the associated latent heat is observable.

4.3.2 *Phase diagram*

With the thermodynamic functions being known, it is obviously possible, with a bit of patience, to work out a complete phase diagram. The entire phase diagram of an SFS trilayer in a relevant region of (T, Λ, D_F) space is mapped out in Fig. 4.4. As we have seen, varying h is equivalent to varying D_F, since results are periodic in D_F because of their periodicity on h, so D_F can be thought of as an experimentally relevant proxy for varying h. The figure shows two views of the same phase diagram: this helps visualize what is really a three dimensional situation. There are three regions in this diagram, each representing one of the three possible states: 0 state, π state, and normal (not superconducting) state. The crossings T_x are calculated from the free energies, and T_c through the linearization method.

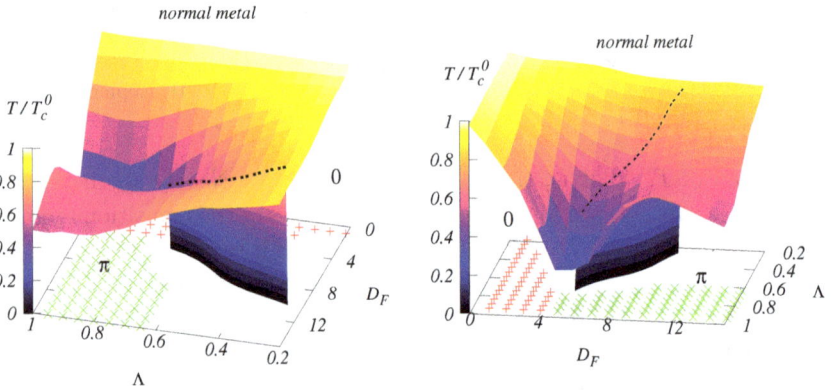

Fig. 4.4 The (Λ, D_F, T) phase diagram for the three-layer system. The two panels show different views of the same plot. There are three regions: in those labeled 0 and π, the 0 and π states are, respectively, the equilibrium state, while *normal metal* indicates where the sample is nonsuperconducting. The top surface separates non-superconducting and superconducting regions. The fairly vertical sheet marks the temperature transitions between 0 and π states. The intersection of the $0 - \pi$ and the T_c boundaries is marked by a dotted line. The portion of the $T = 0$ plane marked by $\times$ symbols is the range of (Λ, D_F) for which only the π state exists for all T: there is no metastable state of the 0 type. Likewise, in the region marked by $+$ symbols only the 0 state exists. In the portion left blank, solutions of both kinds are possible.

The top (higher T) sheet is the superconductor/normal metal transition. As $D_F \to 0$, T_c/T_c^0 approaches unity for all Λ, as it must. At small Λ the sheet also flattens, since then the band width of the ferromagnet is small compared to that in S so that there is, in a sense, less interaction between the Cooper pairs and the ferromagnet. The finite temperature transitions T_x between 0 and π regions are located at the "wall" that goes from the $T = 0$ plane to the T_c sheet, separating the 0 from the π state regions. This wall is of course not completely vertical: its deviation from verticality reflects the first order phase transitions as a function of T. For smaller D_F, this wall ends because one enters a region of parameter space where only one self-consistent solution exists at any temperature. Coincident with this, as one can see more clearly in the right panel, T_c is sharply reduced. In other words, the condensation energies of both states rise towards zero, and one of them actually vanishes. Near this region T_c has always a sharp dip and varies quickly with D_F or Λ. Proceeding towards the opposite end of the wall, at larger values of D_F, T_c increases and the wall becomes steeper, until it eventually becomes vertical. Beyond that, no transition occurs as a function of T: the stable state is the same at all temperatures. Beyond the portion shown, therefore, the wall would become completely vertical. It would be sufficient to show its behavior in the $T = 0$ plane. The crossings at the $T = 0$ plane are not thermodynamic phase transitions, they merely indicate a change in the stable state as various sample parameters are changed. By taking a constant Λ slice of the phase diagram in Fig. 4.4 one can discern again the regular, damped oscillations of T_c with D_F. Figure 4.5 displays T_c for $\Lambda = 0.70$ over a range of D_F. As D_F is increased the amplitude of the oscillations decreases. We shall see below that this agrees with experiment. In addition to T_c, this figure shows $\Delta\mathcal{F}(0)$ for the 0 and π states. In a bulk superconductor, the ratio of this dimensionless normalized quantity to the reduced transition temperature is -0.5, which is confirmed here by the result for the 0 state at $D_F = 0$. The π state is unstable, for obvious reasons, in the $D_F \to 0$ limit. At finite values of D_F this relationship between normalized condensation energy and reduced transition temperature is not strictly obeyed, but there is a qualitative correlation: increases in the absolute value of $\Delta\mathcal{F}(0)$ correspond to increases in T_c. The values of D_F at which the stable state switches between 0 and π correspond to the sharp dips in T_c in all cases. This has also been seen in connection with Fig. 4.4 and it indicates that the structure and shape of the oscillations in T_c are strongly correlated with the low temperature state. This correlation can be understood qualitatively: each switching of the stable state corresponds to a local minimum in the energy, which (at $T = 0$) implies

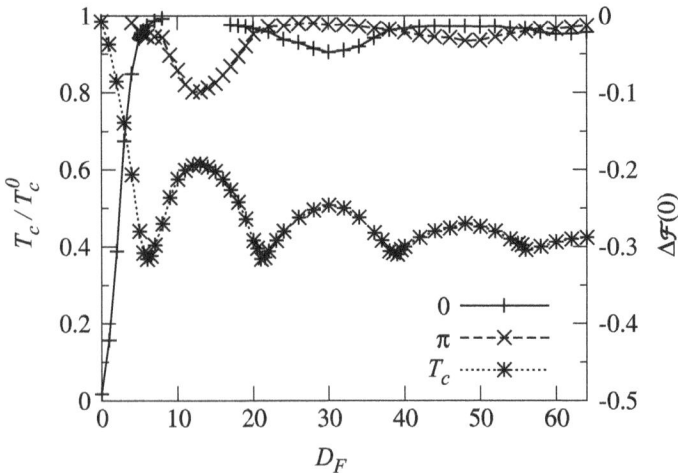

Fig. 4.5 T_c vs. D_F (left scale, $*$ symbols) for $\Lambda = 0.70$. Also shown (right scale) are the energies of the 0 and π states at $T = 0$ (+ and × signs respectively). Notice the correlation between the changes in the stable state at $T = 0$ and the dips in T_c.

a local minimum in the pair amplitude. Near these switching points, the difference in the free energies of the states is a small fraction of the free energy of either of the competing states, and the possibility then arises of the favored state changing with temperature.

The free energy data plotted have gaps, notably for the π state near $D_F \leq 4$ and for the 0 state at $8 \leq D_F \leq 17$. These values of D_F delimit regions in which the self-consistent calculation results in only one state. The free energy of the vanishing state goes continuously to zero at those boundaries. The pair amplitude is found also to go smoothly to zero.

The low temperature crossings at the many different values of D_F suggest the location of more $0 \leftrightarrow \pi$ phase transitions. This is in agreement with direct observations reported in Ref. [13]. Another corroboration of this claim comes from Ref. [14], in which the related parameter which they denote by I_c (the overall critical current of their nonuniform thickness junction) is found to have a significant dip at the $0 \leftrightarrow \pi$ transition temperature. Remarkably, these transitions were observed in Ref. [14] even though their samples did not have layers of uniform thickness. These two experiments and others show that this is an observable and robust phenomenon.

4.3.3 *Experimental comparison for the transition temperature*

It has been known for a long time that oscillations in the transition temperature as a function of geometry are invariably found in experiments,

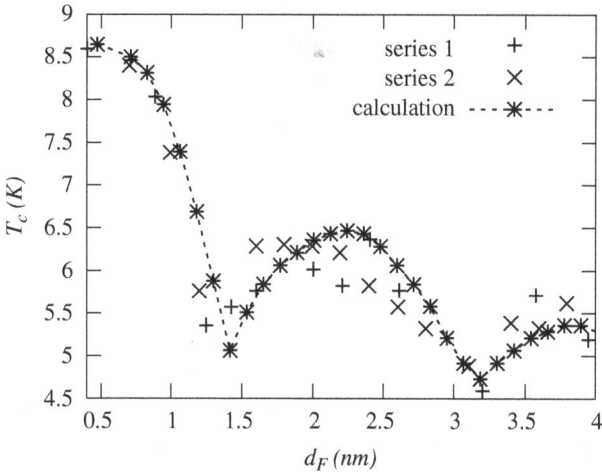

Fig. 4.6 Calculated values of T_c, in good agreement with experimental data for a Nb/Co system. The experimental points shown are the average of the two series reported in Ref. [15].

in hybrid S/F structures. In SFS trilayers, for example, T_c has an oscillatory dependence on d_F. The calculations discussed here also find these oscillations (see Fig. 4.5). The agreement with experiment is furthermore quantitative. In Fig. 4.6, as an example, we see a direct comparison of the theoretical results to classic experimental data for a Nb/Co system of Ref. [15]. In the experiment, the spontaneous magnetization of the Co layer was found to depend on its thickness. This means, in our language, that h is effectively a function of D_F for the purposes of this comparison. To perform the comparison, one simply fits the spontaneous magnetization as reported in Ref. [15] at each thickness, to the elementary expression for m, normalized in terms of the Bohr magneton:

$$m = \left[(1+h)^{3/2} - (1-h)^{3/2}\right] \Big/ \left[(1+h)^{3/2} + (1-h)^{3/2}\right] \qquad (4.6)$$

to extract the effective value of h. In the comparison, a value of $\Lambda = 0.60$ is taken, which is appropriate to the materials mentioned. The experimental and theoretical values are in excellent quantitative agreement on the vertical scale, and the damped oscillations are well aligned in the thickness.

Comparing Figs. 4.6 and 4.5, we can conclude that the dips in Fig. 4.6 should correspond to changes in the stable state at zero temperature. As these changes are, as we have seen, associated with the first order phase transitions, these dips in T_c may also be associated with first order phase transitions, in good agreement with what was reported in Ref. [13]. This implies that studies of T_c are a useful tool for experimental investigation of first order phase transitions and that samples which show dips in T_c are the ones that should be cooled down and studied to locate such phase transitions.

4.3.4 *Reverse proximity effect*

As part of this illustrative example of an SFS trilayer I discuss the reverse proximity effect, that is, the penetration of the magnetization into the S layers and its decrease in the F layer, as it occurs in these trilayers. The value of the local magnetization can be calculated from the wavefunctions in a straightforward way as the difference between the numbers of spin-up and spin-down quasiparticles. Denoting this quantity, in units of the Bohr magneton, as $m(y)$, one has:

$$m(y) = \frac{\langle n_\uparrow(y) \rangle - \langle n_\downarrow(y) \rangle}{\langle n_\uparrow(y) \rangle + \langle n_\downarrow(y) \rangle}, \tag{4.7}$$

where in terms of the wavefunctions obtained as discussed in previous chapters one has:

$$\langle n_\sigma(y) \rangle = \sum_n \left\{ [u_n^\sigma(y)]^2 f(\epsilon_n) + [v_n^\sigma(y)] 2 [1 - f(\epsilon_n)] \right\}. \tag{4.8}$$

The expression Eq. (4.7) correctly reduces to Eq. (4.6) in the bulk F limit.

In general, it is found that there is little difference between $m(y)$ for the 0 and π states. At least at low T, $m(y)$ is dominated by the exchange parameter h and is therefore rather insensitive to the phase of the superconducting state. The value of magnetization induced in the S region is relatively small as, in equilibrium, $m(y)$ decays over the Fermi length scale. To illustrate the effect that temperature has on this trend, we show $m(Y)$ versus the dimensionless length Y at $T = T_x$ in Fig. 4.7. In the figure, the F region is delimited by vertical dotted lines and only a small portion of the S regions is shown. Consistent with what was just mentioned, there is a quick decay and oscillation of $m(Y)$ in the S region. There is a rise in the value of $m(Y)$ to about 0.33 in the center of the F region, which is consistent with the bulk formula below Eq. (4.7) for $h = 0.2$. Indeed, as D_F increases $m(Y)$ flattens to a value that is in good agreement with that estimate. Thus, the local magnetization, while interesting for other reasons, is not a good tool for determining the thermodynamically stable state or locating phase transitions.

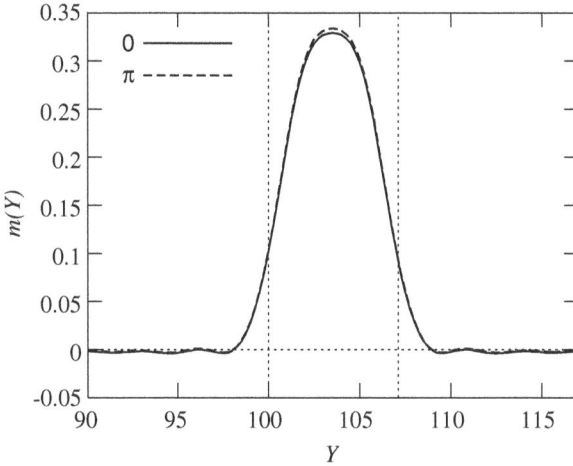

Fig. 4.7 The dimensionless local magnetization $m(Y)$ normalized as explained below Eq. (4.7) for an SFS trilayer at T_x. The parameter values are as in the top panel of Fig. 4.1. Only the central part of the sample is plotted. Results are given for the two nearly coexisting states. The vertical dotted lines delimit the F region.

4.3.5 *The density of states*

An additional equilibrium quantity can be extracted from the self-consistent amplitudes: the energy spectrum. This is the local density of states (LDOS) as a function of energy. In bulk superconductors this is the familiar textbook curve, showing the gap at the Fermi surface and a square root singularity at the gap edge, dropping then to the normal state value which, at the energy scales relevant here, is a constant. It is appropriate to complete this example by showing an example of this quantity for the SFS trilayers. The local density of states is given by

$$N(y, \epsilon) = -\sum_{\sigma}\sum_{n}\left[[u_n^\sigma(y)]^2 f'(\epsilon - \epsilon_n) + [v_n^\sigma(y)]^2 f'(\epsilon + \epsilon_n)\right], \qquad (4.9)$$

where σ denotes spin and $f'(\epsilon)$ is the first derivative of the Fermi function. One can also omit the sum over σ and obtain the spin dependent DOS.

Advanced tunneling spectroscopy techniques are a useful experimental tool to measure the local DOS, thus probing the single-particle spectrum. The local DOS results for 0 and π states are different, including a modified subgap structure. In such cases, tunneling spectroscopy could be used to distinguish the states. We now discuss whether the density of states is also a suitable technique in locating phase transitions.

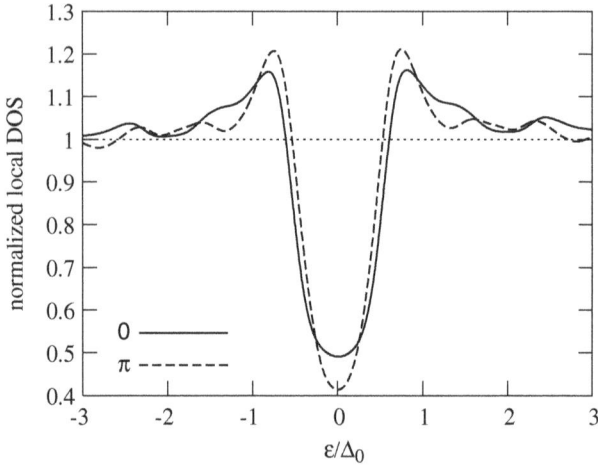

Fig. 4.8 Density of states at T_x for an SFS trilayer. The quantity plotted is the local DOS as defined in Eq. (4.9), averaged over one of the S layers, and normalized to the normal state bulk result in S. The energy is in units of the bulk zero temperature gap in S. The parameter values correspond to the case shown in the bottom left panel of Fig. 4.1 ($T_x/T_c^0 = 0.16$).

In Fig. 4.8 we can see the DOS, defined as the normalized local DOS (from Eq. (4.9)) averaged over one of the S layers, for a typical three layer SFS system (the one considered in the bottom panel of Fig. 4.1) at the temperature where a first order transition occurs with $T_x/T_c^0 = 0.16$. The energy is normalized to the bulk S gap at zero temperature, Δ_0, while the DOS is normalized to its value in a bulk sample of S material in its normal state. In both the nearly coexisting 0 and π states maxima exist near the bulk gap edge, at $\epsilon/\Delta_0 \approx \pm 1$, which, although finite, are qualitatively reminiscent of the divergence found there in a bulk superconductor. The local DOS never quite goes to zero in either state, demonstrating gapless superconductivity: there are bound states in the gap, induced by several scattering processes discussed in Chap. 7. However, the number of states in the gap is clearly larger for the 0 state and the peak structure is markedly lower. Although of course the DOS for both states have some general similarities, the differences that do exist are well within the resolution of current tunneling probes, making the DOS a potentially useful experimental technique in locating the phase transitions or identifying the stable state in the neighborhood of a transition.

With this rather comprehensive example, I conclude the general exposition of the thermodynamic methods.

Chapter 5

Triplet Pairing Generation and Its Consequences

In Chap. 3 and particularly in Sec. 3.4 the conversion of singlet pair amplitudes to triplet was discussed. Their nature and existence was explained in Sec. 3.5 and an example presented in Sec. 3.6.

The existence of triplet amplitudes, and the drastic changes that they induce in the proximity effect is of the most fundamental importance in the thermodynamic and, as we will see later, the transport properties, of the structures we are studying in this book. In this chapter results will be presented for several important and experimentally relevant cases, and effects on the thermodynamic properties will be considered.

As we saw in Secs. 3.2 and 3.3, to create triplet amplitudes without violating conservation laws or the Pauli principle requires special conditions. One possibility is to have, as in Sec. 3.6, more than one magnetic component, with the magnetizations being non-collinear. But this is not the only possibility: one can simply have a single magnet, but with a nonuniform magnetic structure. Or, one can have spin-orbit coupling in the interfacial scattering. In this chapter we will consider both of these cases.

5.1 Triplets in a Simplified Spin Valve Structure

To introduce matters, we consider here the simplified spin valve structure in Fig. 5.1. This is the same as that in Fig. 2.1 except for the lack of an N layer. Note also the different choice for the z-axis: instead of being along the direction of the field in F_1, the two internal fields are aligned at $\pm\alpha/2$ with respect to that axis, as in Fig. 3.1. We already have learned how to

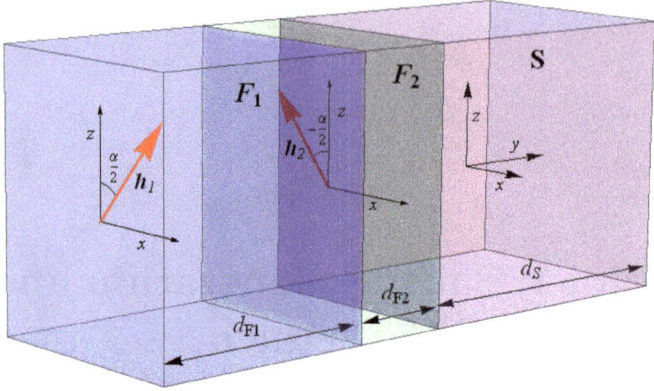

Fig. 5.1 Schematic of the F_1F_2S trilayer considered in this section. The F_1 layer has its internal field oriented at an angle $\alpha/2$ in the $x-z$ plane, while F_2 has orientation at an angle $-\alpha/2$ in the $x-z$ plane.

rotate from one set of axis to another: see Eq. (3.8) and the discussion connected with it.

The results for the triplet amplitudes and the proximity lengths associated with them are obtained using exactly the same procedures as for the FSF case described in Sec. 3.6. But we will begin by presenting results for the transition temperature, obtained by the eigenvalue method described in Sec. 4.2.1. Of course, the matrix elements in Eq. (4.4) differ in many details from those for the SFS structure discussed there. Although Eq. (4.5) is still formally correct, the explicit expression for the matrix elements must be recalculated. The numerical results and the figures presented in this section are from Ref. [16]. In all cases the dimensionless (see Table 2.1) superconducting correlation length is $\Xi_0 = 100$ and all temperatures are again given in units of T_c^0, the transition temperature of *bulk* S material. One has $D_S = 1.5\Xi_0$ unless, as indicated, a larger value is needed to study penetration effects. Except for the transition temperature itself, results shown are in the low temperature limit.

An example of the results for T_c is shown in Fig. 5.2. In this figure both F layers are identical and hence both relatively thin. All three panels in the figure display T_c, normalized to T_c^0, as a function of the angle α. The figure dramatically displays that T_c is not in this case a monotonic function of α. We will see in many other examples that non-monotonic behavior with a misalignment angle is very common, and that it is correlated with singlet to triplet conversion. In the present case, T_c does not in general monotonically

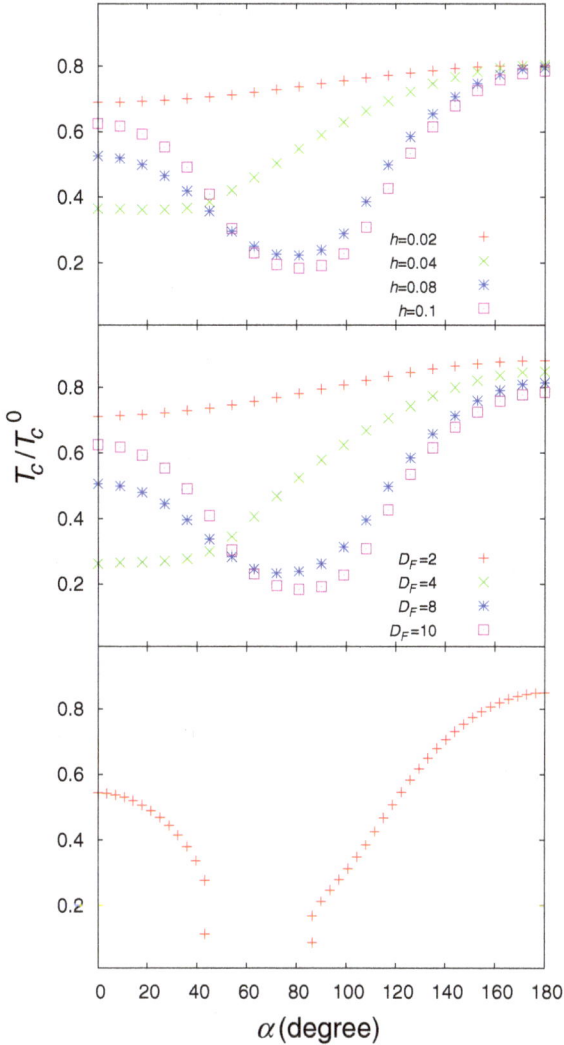

Fig. 5.2   Calculated transition temperatures T_c, normalized to T_c^0. Here the two F layers are identical, $D_{F1} = D_{F2} \equiv D_F$ and $h_1 = h_2 \equiv h$. In the top panel this ratio is shown vs. α for different h values at $D_F = 10$. In the middle panel the same ratio is plotted vs. α for different values of D_F at $h = 0.1$. In the bottom panel T_c vs. α is shown for $D_F = 6$ and $h = 0.15$, a case where reentrance with angle occurs.

increase as α increases from 0 to 180°, but on the contrary it has often a minimum at a value of α typically below 90°.

The top panel, which shows results for several values of h with $D_F = 10$, illustrates the above statements. T_c is in this case monotonic only at the smallest value of h ($h = 0.02$) considered. The non-monotonic behavior starts to set in at around $h = 0.04$ and then it continues, with the minimum T_c remaining at about $\alpha = 80°$. This is not a universal value: for other geometric and material parameters the position of the minimum can be lower or higher. The middle panel has results for a single value of $h = 0.1$ and several values of D_F. This panel makes another important point: the four curves plotted in the top panel and the four ones in this panel correspond to identical values of the product $D_F h$. The results, while not exactly the same, are extremely similar and confirm that the oscillations in T_c are determined by the overall periodicity of the Cooper pair amplitudes in F materials as determined by the difference between up and down Fermi wavevectors, which is approximately proportional (see Chap. 2) to $1/h$ in the range of h shown.

In the lowest panel of the figure we see that reentrance with α can occur in these structures. The results there are for $D_F = 6$ and at $h = 0.15$, a value a little larger than that considered in the other panels. While such reentrance is not the rule, it is not an exceptional situation either: what happens is that the minimum in T_c at intermediate α can simply drop to zero, resulting in reentrance. The origin of this reentrance stems from two combined sources. One is the presence of triplet correlations, which arise from the inhomogeneous magnetization. The other is the usual D_F oscillations in F/S bilayers, that is, the ordinary periodicity of the pair amplitudes introduced in Sec. 2.7.

Having seen this nonmonotonic behavior in T_c it is natural to ask now for the behavior of the free energy. This is shown in Fig. 5.3 for the same system. The quantity plotted is the condensation free energy $\Delta\mathcal{F} = \mathcal{F}_S - \mathcal{F}_N$ from Eq. 4.1, where $\mathcal{F}_S$ and $\mathcal{F}_N$ are the free energies for the superconducting and normal states, respectively. The results are calculated at very low T, hence they are equivalent to the condensation energy. $\Delta\mathcal{F}$ is normalized (as in previous chapters) to $N(0)\Delta_0^2$, where $N(0)$ denotes the density of states at the Fermi level and Δ_0 the bulk value of the singlet pair potential in S: thus (see Ref. [3]) we would have $\Delta\mathcal{F} = -0.5$ for pure bulk S. The three panels in this figure correspond to those in Fig. 5.2. The geometry is the same and the symbol meanings in each panel correspond to the same cases, for ease of comparison. In the top panel, one sees that the $\Delta\mathcal{F}$ curves for $h = 0.02$ and $h = 0.04$ are monotonically decreasing with α. This agrees with the monotonically increasing T_c. One can conclude that as α is changing from $\alpha = 0$ (parallel configuration) to $\alpha = 180°$ (anti-parallel) the superconducting state is becoming increasingly more favorable than

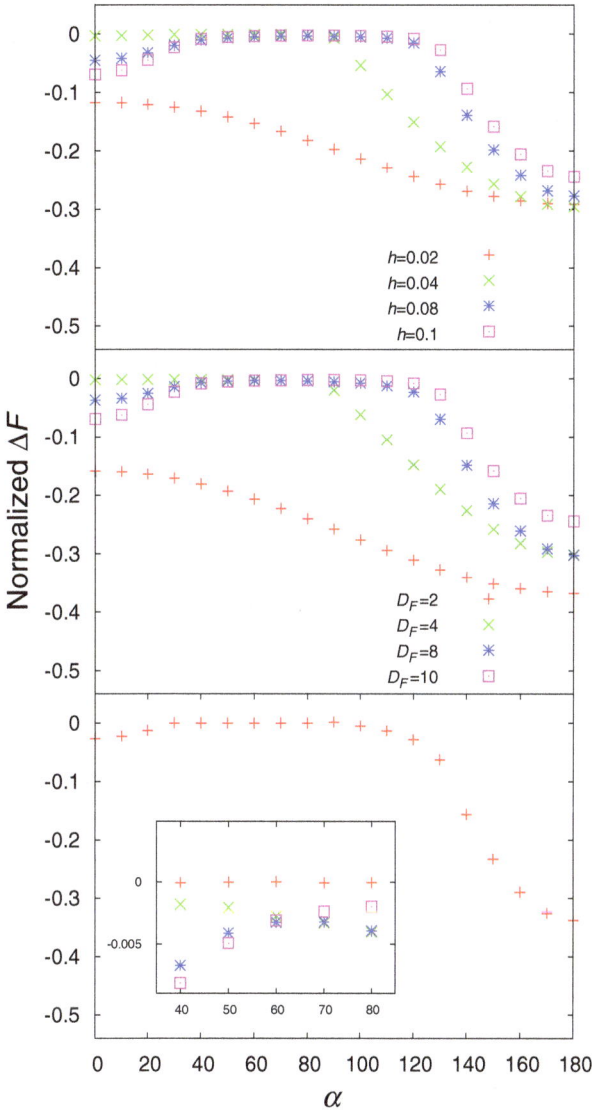

Fig. 5.3 Normalized condensation free energies $\Delta\mathcal{F}$ vs. α, at $T = 0$. The results are for the same geometry and parameter values as in Fig. 5.2, and the symbols have the same meaning. The inset shows the difference between truly reentrant cases and those for which the condensation energy is merely small (see text) in the range of $\alpha = 40°$ to $\alpha = 80°$.

the normal one. The other two curves in this panel, which correspond to $h = 0.08$ and $h = 0.1$, show a maximum near $\alpha = 80°$. Again, this is consistent with the transition temperatures shown in Fig. 5.2. The middle panel of Fig. 5.3 shows $\Delta\mathcal{F}$ for different ferromagnet thicknesses. The curves are very similar to those in the top panel, just as the top two panels in Fig. 5.2 are similar to each other. Therefore, both Figs. 5.2 and 5.3 show that the superconducting states are thermodynamically more stable at $\alpha = 180°$ than in the intermediate regions ($\alpha = 40°$ to $\alpha = 80°$). From the top two panels in Fig. 5.3, we also see that $\Delta\mathcal{F}$ at $\alpha = 180°$ can be near -0.3 in this geometry: this is a very large value, quite comparable to that in pure bulk S. However, in the region of the T_c minima near $\alpha = 80°$, the absolute value of the condensation energy can be over an order of magnitude smaller, although it remains negative. The bottom panel of Fig. 5.3 shows $\Delta\mathcal{F}$ for the reentrant case previously presented in Fig. 5.2, for which $D_F = 6$ and $h = 0.15$. The main plot shows the condensation energy results, which vanish at intermediate angles. Because $\Delta\mathcal{F}$ in the intermediate non-reentrant regions shown in the upper two panels can be very small and hard to see in the vertical scale shown, there is in the lowest panel an inset where the two situations are contrasted using a different scale. In the inset, the plus signs represent $\Delta\mathcal{F}$ for the truly reentrant case and the other three symbols have the same meaning as in the middle panel, where no reentrance occurs. The inset clearly shows the difference: $\Delta\mathcal{F}$ vanishes in the intermediate region only for the reentrant T_c case and remains slightly negative otherwise. The pair amplitudes within the reentrant region can be found [16] self-consistently to be identically zero. Thus one can safely say that in the intermediate region the system must stay in the normal state. That superconductivity in F_1F_2S trilayers can be reentrant with the angle between F_1 and F_2 layers makes these systems ideal candidates for spin-valves. Later in this chapter, we will see an example of reentrant behavior in a bilayer!

Next, we have here some results for the triplet proximity lengths. These results are in terms of the average lengths defined in Eq. (3.9). In Fig. 5.4 we can see results for four lengths thus obtained, for a sample with $D_F = 10$, $D_S = 150$ and $\alpha = 40°$, at several values of h. The left panels show these lengths as extracted from f_0, the real part of $\mathbf{f_0}$, and the right panels show the results for the corresponding f_1 component (see Eq. (3.7)). These quantities are plotted in terms of the dimensionless time τ as defined in Table 2.1. The triplet penetration lengths in the F region become independent of τ beyond small values of this quantity, for both f_0 and f_1.

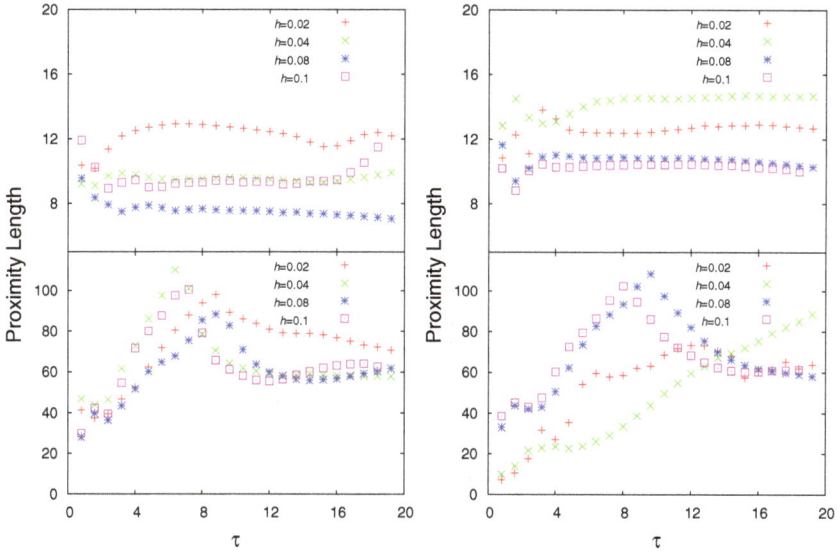

Fig. 5.4 Study of the triplet proximity lengths, see Eq. (3.9), vs. τ. For this figure, $D_F = 10$, $D_S = 150$, and $\alpha = 40°$. Left side: lengths as extracted from $f_0(Y)$ for several values of h in the F regions (top) and in the S region (bottom). The right side contain the same information, arranged in the same way, but with the proximity length extracted from $f_1(Y)$. The lengths become independent of τ.

The triplet correlations easily pervade the entire magnetic part of the sample. The corresponding penetration lengths in the S region are in both cases substantially greater which reflects that D_S is much larger. They do become τ-independent, but before they do there is a peak around $\tau = 8$ in all cases except for f_1 at larger h where it is beyond the figure range.

5.2 Spin Active Interfaces

There is of course another rather obvious way to generate triplet correlations and this is to have spin orbit scattering. The easiest and physically most relevant way is to assume that interfacial scattering is spin dependent, that is, that the interfaces are "spin active". Then, the conservation laws allow the singlets to change to triplets, and only one F layer is needed. Here we will discuss only a very simple case: an SFS trilayer with spin active interfaces and (to make the results more salient) a half-metallic F layer: $h = 1$. The results in the figures in this section are from Ref. [20].

The presence of spin active interfacial scattering requires a slight modification of Eqs. (2.1) and (2.2) and, of course, Eq. (2.10). Briefly, we can write

$$\mathcal{H}_{\text{eff}} = \int d^3r \left\{ \sum_\alpha \psi_\alpha^\dagger(\mathbf{r}) \left[-\frac{\boldsymbol{\nabla}^2}{2m} - E_F + V_0(\mathbf{r}) \right] \psi_\alpha(\mathbf{r}) \right.$$

$$+ \sum_{\alpha,\beta} \psi_\alpha^\dagger(\mathbf{r})(\mathbf{V} \cdot \boldsymbol{\sigma})_{\alpha\beta} \, \psi_\beta(\mathbf{r}) + \frac{1}{2} \left[\sum_{\alpha,\beta} (i\sigma_y)_{\alpha\beta} \Delta(\mathbf{r}) \psi_\alpha^\dagger(\mathbf{r}) \psi_\beta^\dagger(\mathbf{r}) + h.c. \right]$$

$$+ \left. \sum_{\alpha,\beta} \psi_\alpha^\dagger(\mathbf{r})(\mathbf{h} \cdot \boldsymbol{\sigma})_{\alpha\beta} \, \psi_\beta(\mathbf{r}) \right\}. \qquad (5.1)$$

The meaning of the symbols is as in Chap. 2. The important difference is the spin flip scattering term, confined to the two interfaces at $y = d_S$ and $y = d_S + d_F$. It takes place in the $x - z$ plane:

$$\mathbf{V} \cdot \boldsymbol{\sigma} = V_x(y)\sigma_x + V_z(y)\sigma_z. \qquad (5.2)$$

The V_z component represents an unimportant local modification of the h field, while V_x is the spin flip term. One can take $V_y = 0$ because of the layered geometry. Each of the triplet states can potentially exist over large length scales, thus allowing competing orderings to coexist.

One then proceeds as in Chap. 2, going to the quasi one-dimensional limit and repeatedly diagonalizing to obtain self-consistency. The only difference arises from the interfacial scattering. To begin with, the interfacial potential, instead of having been given by Eq. (2.10) is now given by:

$$V_i(y) = V_i[\delta(y - d_S) + \delta(y - (d_S + d_F))], \qquad (5.3)$$

where as explained above, $i = x, z$. The dimensionless parameter that characterizes the spin-orbit scattering at the interface can be taken to be $H_{spin} \equiv 2mV_x/k_F$, which except for a factor of two is the normalization chosen for the spin independent part in Table 2.1. The presence of the spin flip x-component, $V_x(y)$, produces some small technical computational complications which are however on the same scale as those that arise from the presence of non-collinear fields.

Let us then see some results. The SFS system to which the figures below correspond has a thickness d_S such that $D_S \equiv k_F d_S = 300 = 6Xi_0$ so that d_S considerably exceeds the superconducting coherence length.

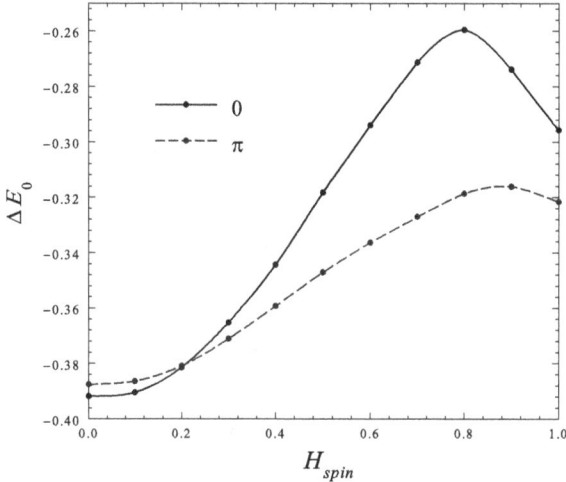

Fig. 5.5 Normalized condensation energy ($\Delta \mathcal{F}$ at $T = 0$) vs. the dimensionless parameter H_{spin} characterizing the spin flip strength (see text) at the interfaces. Results are for an SFS trilayer with a half metallic ferromagnet. We see that for this case both the π and the 0 states are locally stable for all values of H_{spin} included and that the π state has lower condensation energy except at small values of H_{spin}.

The two S layers are separated by a ferromagnetic layer, which must be taken to be thin enough so that the two superconductors are still coupled through the F material via the proximity effect. In this example $D_F = 10$. All results shown are, as usual, in the low temperature limit.

For the parameter values considered, there are two local minima of the free energy (that is, two self-consistent solutions to the BdG equations). These correspond, of course, to 0 and π states, these being, we recall, the values of the phase difference between the two S layers. As we have seen in Chap. 4 one must then find out which of these two local minima is the absolute minimum, via Eq. (4.1). At zero temperature, this is the condensation energy. The results depend on H_{spin} as shown in Fig. 5.5. We see in the figure that, as in Chap. 4, the condensation energies are reduced, in all cases, from what they would be for a bulk S sample. For all values of H_{spin} both the 0 and π configurations of the structure are at least locally stable, but in general non-degenerate, showing that indeed the two S slabs are indeed coupled via the proximity effect. At very small values of the spin flip parameter, the 0 configuration is the stable one, but this changes as H_{spin} increases: there is a first order phase transition

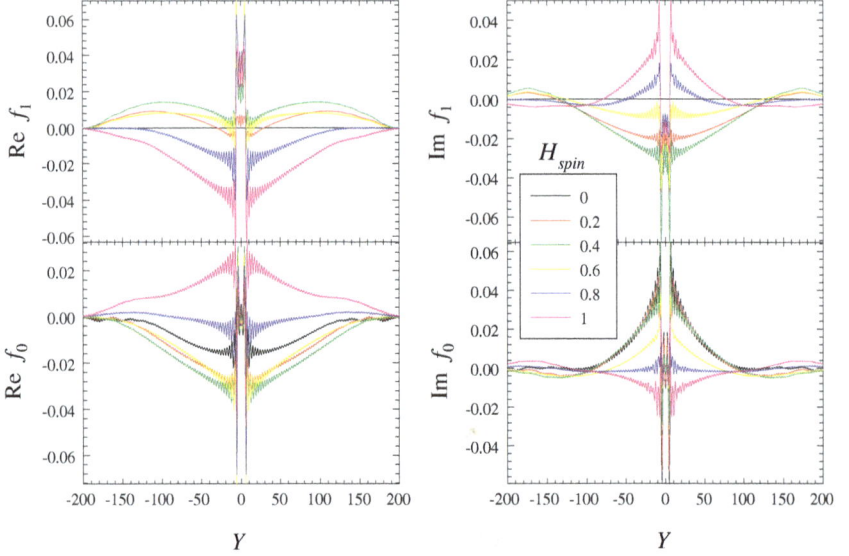

Fig. 5.6 The $\mathbf{f_1}$ and $\mathbf{f_0}$ triplet pair amplitudes (Eq. (3.7)) for a 0-state SFS junction plotted as a function of Y for several values of H_{spin} as indicated in the legend. The left panels show the real part while the right ones show the imaginary part. The F layer is at $-5 < Y < 5$. Geometry and parameters are as in the previous figure.

at $H_{spin} \lesssim 0.2$ and at larger values of H_{spin} the π configuration is the stable one and the 0 configuration is much less stable. The 0 configuration metastable minimum is shallowest at $H_{spin} \approx 0.8$ where the condensation energy has a sharp maximum. The condensation energy of the π state is more weakly dependent on H_{spin} with a maximum near $H_{spin} = 0.82$ much shallower than that in the 0 state.

Next, we discuss briefly the spatial dependence of the complex triplet pairing functions $\mathbf{f_0}(Y, \tau)$ and $\mathbf{f_1}(Y, \tau)$ as defined in Eq. (3.7). In Fig. 5.6 there are results for the Y dependence of the triplet amplitudes for the 0 solution (although this is usually not the stable one, see Fig. 5.5). The results plotted are for fixed $\tau = 20$, and six values of H_{spin} in the range $0 \leq H_{spin} \leq 1$ (see legend). All amplitudes are normalized so that the bulk triplet amplitude would be unity. At this value of τ the triplet pairing states have penetrated most of the two S regions. The range of Y included in the plots is, for clarity, somewhat narrower than the sample size: regions where the amplitudes are very small or zero are omitted. One can see that the amplitude $\mathbf{f_1}$ vanishes identically in the absence of spin-flip scattering, since in that case both the total spin and its z component are good quantum numbers. For finite values of H_{spin}, all possible projections of the total spin exist.

Turning our attention to the real part of $\mathbf{f_1}$ as plotted we see that it shows a monotonic decline in magnitude from the interface, over about three to four coherence lengths, for the largest two spin flip strengths (with a superposition of rapid oscillations). The remaining weaker scattering strengths are quite different: they yield nonmonotonic behavior with a maximum located deep into S at about $2\xi_0$, and then decaying to zero at roughly $4\xi_0$. These amplitudes, $\mathrm{Re}f_1$, are predominately positive for higher spin transparency (smaller H_{spin}) junctions, and then undergo a sign flip for the stronger H_{spin}. If we examine now the imaginary component $\mathrm{Im}f_1$, we see similar opposite parity effects separating the strongest H_{spin} from the weaker values. Here, however, there is a clearer separation between the curves. The $m = \pm 1$ triplet correlations have reached deeper within the sample for the same τ.

Similar plots can be made [16] for the π state. Overall, the π state f_1 amplitudes are suppressed, even for the case when both the 0 and π states have the same condensation energy ($H_{spin} \approx 0.2$). The diminished π state results arise partly from the symmetry requirements imposed upon f_i: $f_i(-Y) = -f_i(Y)$, thus in F, the f_i amplitudes vanish at $Y = 0$, which can constrain the overall longer range spatial behavior. The singlet Cooper pair order parameter that minimizes the free energy, and for which $|\Delta(Y)|$ is typically larger, does not necessarily result in the larger triplet amplitudes.

The triplet amplitude, $\mathbf{f_0}$, with zero projection of the z-component of total spin, does not vanish, at any finite time, even for $H_{spin} = 0$, since as we have already discussed, the total spin (as opposed to its z component) is not a good quantum number in the presence of the F material. It does still vanish at $t = 0$ because of the Pauli principle. Thus, in general, when a single quantization axis exists for the system, f_0 coexists with the ordinary singlet s-wave component. The results for $\mathbf{f_0}$ are different from those for $\mathbf{f_1}$ since the former does not emerge solely from H_{spin}. There are two competing spin flip effects in the z direction: the magnetization of the half metal, and the spin dependent scattering at the interface. It is clear from Eqs. 5.1 and 5.2 that these two effects compete against each other. The spatial symmetry of the singlet Cooper pair is also reflected in the triplet pairing states: if the singlet order parameter is in a 0 junction state, the corresponding triplet amplitudes maintain that symmetry.

5.3 Nonuniform Magnets: Triplets and Reentrance

There is yet another way to generate triplet pairing with only one F layer, and that is by using a nonuniform magnet. Experimentally, Holmium has been used [17] to make Nb–Ho bilayers. Holmium is an antiferromagnet below 133 K and transforms itself into a spiral conical ferromagnet at about 20 K. Because of the non-uniformity of the magnetic structure, triplet conversion is allowed. In this section, we will see how triplet generation in $S/(\text{spiral } F)$ bilayers leads to very peculiar phenomena, including an oscillatory behavior for T_c which may lead to a reentrant phase diagram.

A sketch of the structure considered is in Fig. 5.7. The internal field in the F layer is given by:

$$h = h_0 \left\{ \cos\alpha\hat{\mathbf{y}} + \sin\alpha \left[\sin\left(\frac{\beta y}{a}\right)\hat{\mathbf{x}} + \cos\left(\frac{\beta y}{a}\right)\hat{\mathbf{z}} \right] \right\}, \qquad (5.4)$$

where the helical magnetic structure has a turning angle β, and opening angle α. For Ho, one has $\alpha = \pi/6$ and $\beta = 4\pi/9$. These are the values used in the results [18, 19] presented in this section, although the amplitude will be varied. In these results, a, the lattice constant, is taken as the unit of length. The spatial period of the helix is $\lambda_0 = 2\pi a/\beta$.

Taking into account that the field has three nonzero components, the eigenvalue equation in this geometry is a bit more complicated than in

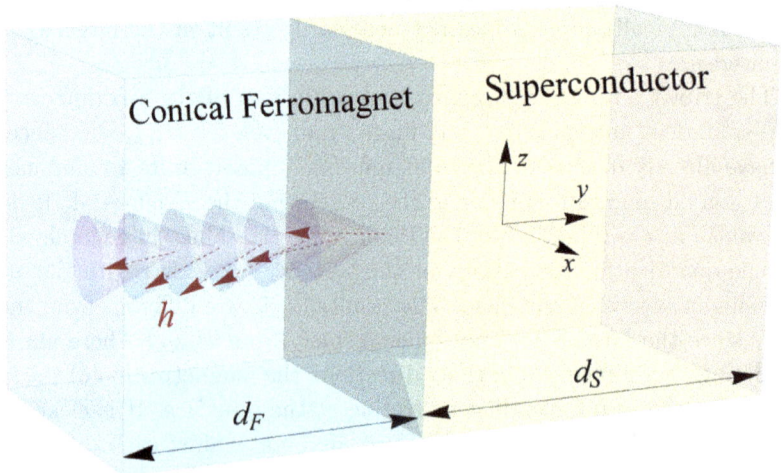

Fig. 5.7 Diagram of a conical ferromagnet-superconductor bilayer. The spiral magnetic structure is described by an exchange field h as in Eq. (5.4). The notation for the relevant widths is indicated.

previous cases. It takes the form:

$$
\begin{pmatrix}
\mathcal{H}_0 - h_z & -h_x + ih_y & 0 & \Delta(y) \\
-h_x - ih_y & \mathcal{H}_0 + h_z & -\Delta(y) & 0 \\
0 & -\Delta(y)^* & -(\mathcal{H}_0 + h_z) & h_x + ih_y \\
\Delta(y)^* & 0 & h_x - ih_y & -(\mathcal{H}_0 - h_z)
\end{pmatrix}
$$

$$
\times
\begin{pmatrix}
u_{n\uparrow}(y) \\
u_{n\downarrow}(y) \\
v_{n\uparrow}(y) \\
v_{n\downarrow}(y)
\end{pmatrix}
= \epsilon_n
\begin{pmatrix}
u_{n\uparrow}(y) \\
u_{n\downarrow}(y) \\
v_{n\uparrow}(y) \\
v_{n\downarrow}(y)
\end{pmatrix}.
\tag{5.5}
$$

These equations are solved self-consistently in the usual way. Once self-consistency is obtained, one can compute all relevant quantities including the ordinary singlet amplitude, $F(y)$, ($\Delta(y) = gF(y)$), the triplet amplitudes, as given from Eq. (3.7), and the free energy from Eq. (4.1). It is also, as we shall see, very interesting to compute T_c, which is done via the linearization method of Sec. 4.2.1.

5.3.1 *Results for the spiral ferromagnet/superconductor bilayers*

There are two things that lead to interesting results for this system. One is that there is triplet generation without the need to have more than one F layer, or a spin-active scattering. The other is that the additional periodicity due to the conical structure interacts with the usual periodicity due to the different Fermi surface wavevectors, leading to new phenomena. All results in this subsection use the same dimensionless conventions as in Table 2.1 *except* that, as mentioned, the lengths are in units of a. The correlation length is fixed to 100 such units. The wavelength of the magnetic structure (see below Eq. (5.4)) is twelve lattice constants. The usual characteristic oscillation lengths, as introduced in Sec. 2.7, go as $1/h$. The figures are taken from [18, 19] except those in Sec. 5.3.1.1 which are from Ref. [17].

Consider the results for T_c, as usual normalized to T_c^0, shown in Fig. 5.8 The D_F range includes three complete periods of the spiral magnetic order. This is vividly reflected in the results shown: indeed, the presence of these oscillations is the most prominent feature in the figure. The oscillations in T_c arise (as we discuss below) from a combination of the periodicity of the spiral magnetic structure and the usual T_c oscillations which occur, even when the magnet is uniform, from the difference in the wavevectors of the up and down spins. A range of values of h is included. For Ho,

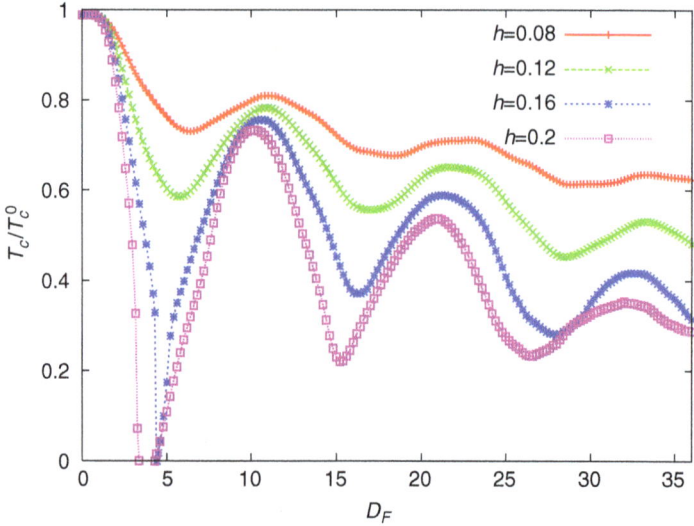

Fig. 5.8 Calculated transition temperatures T_c, normalized to T_c^0, for an FS bilayer with spiral magnetic structure for F. It is plotted vs. D_F for several values of the dimensionless exchange field h. In this figure D_S is fixed to $1.5\Xi_0$. The lines connecting data points are guides to the eye.

$h \approx 0.1$. In Fig. 5.8, one can also see that with stronger exchange fields the oscillation amplitudes are larger. Despite this increase of the amplitudes with the exchange field (they are approximately proportional to h), the overall T_c decreases when the exchange field increases. A stronger exchange field destroys the superconductivity more efficiently.

Another feature seen in this figure is the decrease of the amplitude oscillations with increasing D_F. This arises simply because the singlet Cooper pair amplitudes in S near the F/S boundary decay more strongly at larger d_F and therefore the effect of the pair amplitude oscillations in F is weaker. In a F/S bilayer where the ferromagnet is homogeneous, the periodicity of the T_c oscillations is governed only by the exchange field. Here, where a bilayer with a conical inhomogeneous ferromagnet is considered, the intrinsic spiral magnetic order with spatial period $\Lambda_0 \equiv \lambda_0/a = 2\pi/\beta$ (see below Eq. (5.4)) plays an equally important role in the T_c oscillations, competing with that of $1/h$. (Please do not confuse this Λ_0 with the mismatch parameter of Table 2.1). In other words, both the strength and the periodicity of exchange fields influence the overall decay and the oscillatory nature of the superconducting transition temperatures. The existence of two different spatial periodicities leads to the obvious consequence that the $T_c(D_F)$ curves are not describable in terms of one single period. However, when h is not very strong ($h \lesssim 0.1$), the minima of T_c are near the locations where $D_F = \Lambda_0/2$, $3\Lambda_0/2$, and $5\Lambda_0/2$. Similarly, the T_c maxima occur near $D_F = \Lambda_0$, $2\Lambda_0$, and $3\lambda_0$. This indicates that the magnetic periodicity is dominant. As h increases deviations become obvious. Figure 5.8 shows that the distances between two successive maxima decrease when the exchange fields increase.

The depth of the minima in T_c is considerable and indeed the system goes normal in some cases, particularly when the exchange field is strong enough: the systems can become normal in some range $D_{F1} < D_F < D_{F2}$. Indeed, reentrance with D_F can be seen to occur in Fig. 5.8 near $D_F = 4$ at $h = 0.2$.

This reentrance with geometry makes one wonder if reentrance with temperature may in some cases occur. And the answer is definitely yes. An example is given in Fig. 5.9. These are results for $h = 0.15$. As explained in

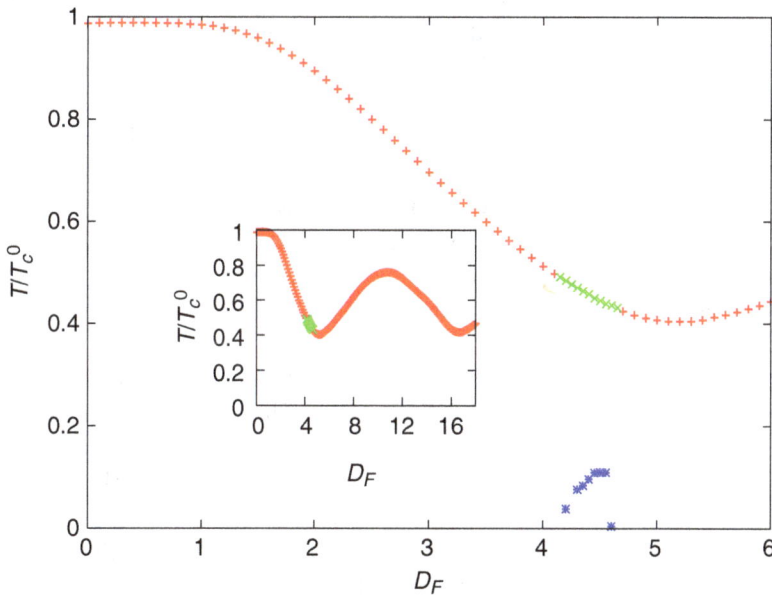

Fig. 5.9 Normalized transition temperature at $h = 0.15$ vs. the dimensionless ferromagnet width, $D_F(\equiv d_F/a)$. Main plot: The upper points ($+$ and $\times$ signs) are the usual critical temperature (T_{c2}), leading to the superconducting state as T is lowered. In the region $4 \lesssim D_F \lesssim 5$ (highlighted by the different $\times$ signs) a second transition back to the normal state appears at the star points forming the lower "dome". The inset shows a broader range of magnet widths, revealing the overall periodicity of T_{c2}.

Sec. 4.2.1 one can calculate the transition temperature either by lowering T until the largest eigenvalue of the matrix in Eq. (4.4) increases to unity or by starting at low T and increasing it until the largest eigenvalue decreases to unity. Ordinarily, both methods give the same result, and doing it both ways is merely a check on the numerical accuracy of the procedure. But in this region things are more complicated: in the usual case the largest eigenvalue is smaller than unity when T is larger than T_c, while in a reentrant case with superconductivity in the range $T_{c1} < T < T_{c2}$ (but not at $T < T_{c1}$ or $T > T_{c2}$) one finds that for $T < T_{c1}$ the largest eigenvalue is less than unity and that upon increasing T from zero this eigenvalue reaches unity at $T = T_{c1}$. On the other hand, one finds T_{c2} by *decreasing* T from $T > T_{c2}$ until the largest eigenvalue reaches one.

This reentrant behavior is far from unique to the case shown in Fig. 5.9. There are other regions in the d_S, h, and d_F parameter space where this reentrance with T occurs: superconductivity exists only in a temperature range $T_{c1} < T < T_{c2}$, where T_{c1} is finite. It appears that locations near minima in the $T_c(D_F)$ curve are favorable for such an occurrence. Physically, this is because superconductivity is relatively weak near these minima. This situation can be viewed as a "compromise": near a minimum of the main T_c curve, superconductivity is not completely destroyed on lowering T but it becomes "fragile" and can disappear upon further lowering T.

To understand the reentrance phenomena in T it is most useful to examine the behavior of the ordinary, singlet, pair amplitude and also the thermodynamics at the two transitions and in the region between them. Given the condensation free energy $\Delta\mathcal{F}$, evaluated from Eq. (4.1), other quantities such as the condensation energy and entropy are easily obtained.

It is illuminating to consider first the T dependence of the singlet pair amplitude $F(Y)$ well inside the S material. This quantity, normalized to its value in bulk S material, is plotted in Fig. 5.10 (main plot, left scale) as a function of T. The behavior of the pair amplitude in the reentrant region is quite striking. There are two transition temperatures: below a very low but finite temperature, $T_{c1}/T_c^0 = 0.07$, the singlet pair amplitude vanishes and the system is in its normal state. The singlet amplitude then begins to rise continuously, has a maximum at a temperature T_m, (where $T_m/T_c^0 \approx 0.3$) and eventually drops to zero, again continuously, at an upper transition $T_{c2}/T_c^0 \approx 0.48$. In the region $T_{c1} < T < T_{c2}$ the system is in the superconducting state. Both transitions are of second order. The values of T_{c1} and T_{c2} from the vanishing of the amplitude, as seen in Fig. 5.10, agree

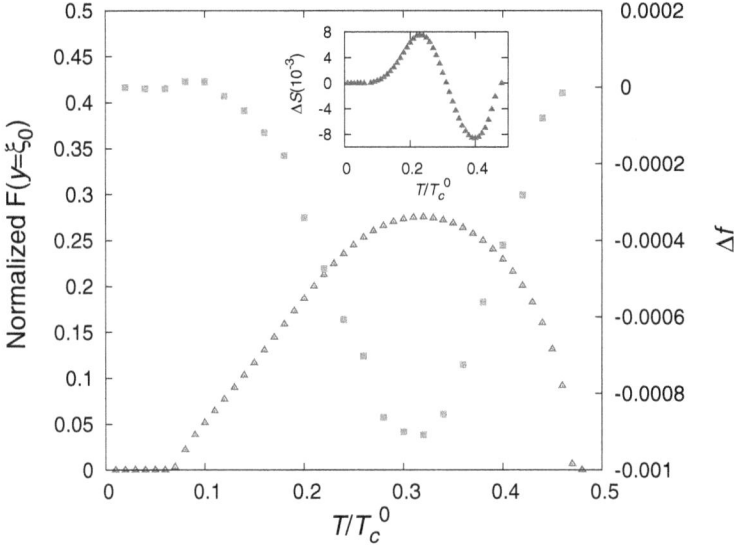

Fig. 5.10 Pair amplitude and thermodynamic functions in a spiral FS bilayer plotted vs. T/T_c^0. In the main plot, the triangles and left vertical scale display the normalized singlet Cooper pair amplitude $F(Y)$, one correlation length inside S. This quantity vanishes at the upper transition temperature ($T_{c2} \approx 0.47 T_c^0$) and again at the lower transition $T_{c1} \approx 0.07 T_c^0$. The squares and right scale are the normalized condensation free energy, $\Delta\mathcal{F}$. The vanishing of $\Delta\mathcal{F}$ at the upper and lower transitions is clearly seen. The inset shows the normalized entropy difference $\Delta S \equiv -(d\Delta\mathcal{F}/d(T/T_c^0))$.

with those calculated directly from linearization of the eigenvalue equation, as plotted in Fig. 5.9.

We now turn to the condensation free energy, $\Delta\mathcal{F}$, and entropy, ΔS, for the same T reentrant case. $\Delta\mathcal{F}$ is shown in the main panel (and right scale) of Fig. 5.10. The inset shows the normalized condensation entropy, defined in Eq. (4.2). When the system is near (but outside) the reentrant region, the behavior of both quantities plotted is qualitatively similar to that found in textbooks for bulk BCS superconductors. Quantitatively, the magnitude of $\Delta\mathcal{F}$ for these systems is much smaller than that for bulk S where we would have $\Delta\mathcal{F} = -0.5$ at $T = 0$ in our units. Moreover, the vanishing of the entropy difference at a finite T_c confirms the occurrence of a second order phase transition. The value of this transition temperature agrees with all above results.

Although the values of $\Delta\mathcal{F}$ are small, one can still find that the minimum of $\Delta\mathcal{F}$ occurs at approximately the same value T_m where the singlet pair amplitudes have a maximum. Thus, the superconductivity is most robust at $T = T_m$. The two transition temperatures T_{c1} and T_{c2} can also be determined from the free energy in Fig. 5.10 and match those found in Fig. 5.9 or from the pair amplitude. In the two T ranges $T < T_{c1}$ and $T > T_{c2}$, the normal state is the only self-consistent solution to the basic equations. The vanishing $\Delta \mathcal{F}$ when $T < T_{c1}$ means that the electrons do not then condensate into Cooper pairs.

There are some remarkable facts about the behavior of ΔS in the reentrant case. ΔS is *positive* for $T_{c1} < T < T_m$ where T_m is again the value of T at which the singlet pair amplitude reaches its maximum *and* $\Delta \mathcal{F}$ its minimum. That the entropy of the superconducting state is higher than that of the normal state indicates that the normal state at $T_{c1} < T < T_m$ is *more* ordered than the superconducting one. This truly unusual fact, which is the root cause of the reentrance, is due to the oscillating nature of both the Cooper pair condensates and of the exchange field, which leads to an uncommonly complicated structure for the pair amplitude. Above T_m, the superconducting state becomes more ordered than the normal state: hence ΔS is negative.

Let us now have a look at the triplet amplitudes in this case. We do this in Fig. 5.11. The decay of these amplitudes can be again characterized by the proximity lengths defined in Eq. (3.9). These lengths depend on D_S, D_F, I, and τ. The range of D_F considered here, see Fig. 5.11, is from Λ_0 to $3\Lambda_0$. In this figure these proximity lengths on both the F and S sides are plotted versus D_F for three different values of h, at $\tau = 4.0$. The left panels show the proximity lengths as extracted from the real part of $\mathbf{f_0}$ and the right panels those similarly extracted from $\mathbf{f_1}$. Let us examine first the F side (top two panels). One can clearly see that both $\ell_{0,F}$ and $\ell_{1,F}$ are correlated with the strength of the exchange fields. The results display a period of near $\Lambda_0/2$ for both $\ell_{0,F}$ and $\ell_{1,F}$ at $h = 1.0$. The peak heights increase only slowly with increasing D_F. Also, the locations of the maxima or minima of $\ell_{0,F}$ are locations of minima or maxima, respectively, of $\ell_{1,F}$. This is as one might expect from the rotating character of the field. On the other hand, for $h = 0.1$ or $h = 0.5$ the periodicity is not clear since, for reasons already mentioned, the intermingling of periodicities becomes more complicated. Overall the proximity lengths are larger than those in the half-metallic limit. However, one can still say that both gradually increase, although with fluctuations, with D_F.

The results for the superconducting side are as plotted in the bottom panels of Fig. 5.11. The minimum of $\ell_{0,S}$ is, for all three values of h, at

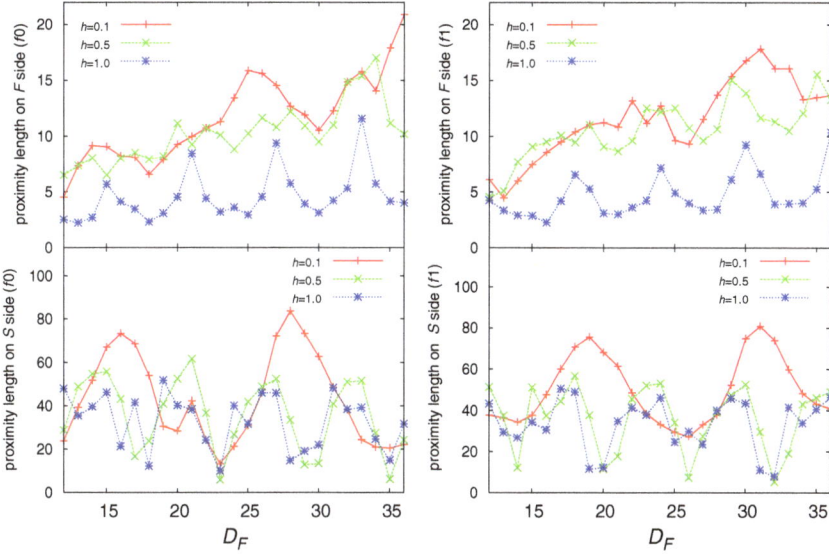

Fig. 5.11 The proximity lengths (see Eq. (3.9)) for the FS bilayer problem in previous figures, of the induced triplet pair amplitudes vs. D_F for different h (legend). The left panels show the proximity lengths obtained from the real parts of $\mathbf{f_0}$ in the F and S regions, and the right panels those extracted from $\mathbf{f_1}$. The lines are guides to the eye.

$D_F = 23$. This value is near $2\Lambda_0$. The maxima of $\ell_{0,S}$ for $h = 0.1$ are at $D_F = 16$ and $D_F = 28$ which are not far from $1.5\Lambda_0$ and $2.5\Lambda_0$ respectively. On the other hand, for $\ell_{1,S}$ the maxima for $h = 0.1$ are at $D_F = 19$ and $D_F = 31$ and there is a minimum at $D_F = 26$. The locations of these maxima are still near $1.5\Lambda_0$ and $2.5\Lambda_0$ and they are only slightly different than what they are for the $m = 0$ case. Recall that (Fig. 5.8) the maxima of T_c occur when D_F is close to an integer multiple of Λ_0: since a higher T_c is correlated with a higher singlet pair amplitude, this suggests again that there exists a conversion between singlet and triplet Cooper pairs. The dependence on D_F on the S side, at $h = 1.0$, is harder to characterize. This is because the high value of h reduces the scale of the overall proximity effect in S (i.e., the depletion of the singlet amplitude). At $h = 0.5$, one still finds that the approximate periodicity of both lengths is about $\Lambda_0/2$. The maxima and minima of the two lengths are again anti-correlated at $h = 0.5$.

Let us look now briefly at some experimental results for these bilayers. In Chap. 6 a more comprehensive comparison between theory and experiment for a more complicated system will be given.

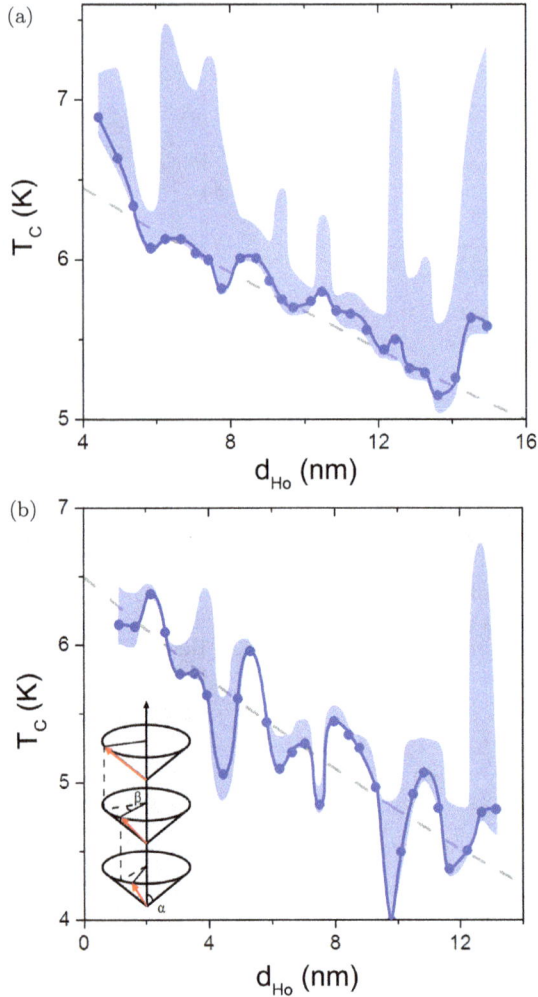

Fig. 5.12 Experimental transition temperatures for two Ho/Nb bilayers. Results are for two different sample series, fabricated separately and labeled by the Nb thickness in nanometers [17]. The oscillations are superimposed to an exponential decay (dotted lines).

5.3.1.1 *Experiments in Ho/Nb bilayers*

As already mentioned, experiments have been performed on good Ho/Nb bilayers. Here we look at some results [17] for T_c, as plotted in Fig. 5.12 and compare with theory in Fig. 5.13.

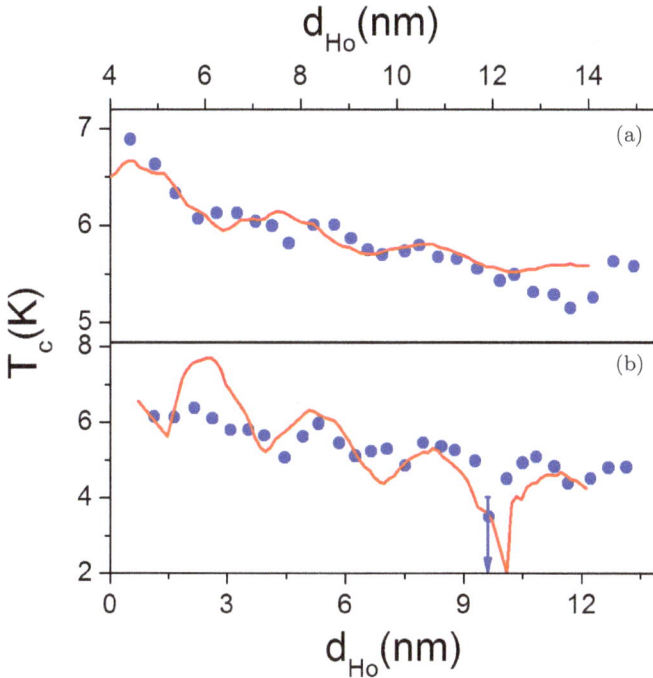

Fig. 5.13 Calculated (see text) transition temperatures for the same two Ho/Nb bilayers as in Fig. 5.12.

Two sample series were fabricated, differing in Nb layer thickness. The experimental results for T_c are plotted vs. the Ho thickness in Fig. 5.12. The two Nb thicknesses studied are indicated in the labels. Note the extremely strong thickness dependence of the results, in agreement with theory. On an overall general decrease of T_c with d_{Ho} one sees strong oscillations (particularly in the thinner sample) that include a deep minimum near $d_{Ho} = 9.6$ nm.

In Fig. 5.13 we can see the theoretical results, obtained by the methods discussed above, to the experimental data. A value of $T_c^0 = 9$ K was used (consistent with extrapolating the experimental curves to zero thickness) and standard values [17] were used for the Nb superconducting coherence length and Fermi speed. The results are, as just mentioned, very sensitive to the value of d_s, therefore the agreement is quite remarkable. Slightly different values of h_0 in the vicinity of 0.14 were used for best fit. For both series, exponential decay with superimposed oscillations was obtained.

The very deep minimum near 10 nm for the thinner sample reveals the proximity of reentrance. The theoretical results reveal that the system is then in close proximity to reentrance.

5.4 Summary

We have seen in this chapter that there are many different ways that the conservation and Pauli principle strictures regarding singlet to triplet conversion, discussed in Chap. 3, can be refined. In addition to those discussed here there are others: incorporating a multi-domain magnet is an example. What is remarkable is that all of these methods are efficient: once triplets are *allowed* they sprout in reasonable quantity. This is not at all obvious *a priori*. It could have happened that triplets would not form, or do so in negligible amounts. The only argument I can think of is that enlarging parameter space will in general lead to lower allowed free energy minima, but this is not a quantitative argument.

Chapter 6

Experiments: Transition Temperature and Triplet Conversion in Spin Valve Structures

Although this is a theoretical book, it is only proper to connect with experiment in the right places. For example, in Sec. 4.3.3 in Chap. 4 we saw a brief example of how equilibrium data for the transition temperature can be compared with experiment in a simple system, and we have just seen in Sec. 5.3.1.1 how to do so for Ho/Nb bilayers. But this chapter will be different: in it we will see a very detailed quantitative comparison for an extensive series of experiments. In addition, we will include a *quantitative* study of the singlet to triplet conversion, based on the results of the same experiments.

The experiments [21] which we discuss in this chapter were done in a multilayer with an SF_1NF_2 structure. For reasons that will appear clearer later in this book (see e.g., Sec. 8.4.4) this is called, as already done in Sec. 2.1, a spin-valve configuration. As explained in Chap. 3 it is necessary to have more than one F layer for the singlet to triplet conversion to be possible: that is, so that the conservation laws and the Pauli principle are not violated. Since the conversion, we recall, can take place only when the internal fields are non-collinear, it is also very important to have the ability to manipulate the angle between them. A moment's thought will reveal that this ability requires the two F layers to be physically separate: there has to be an interface between them so that the magnetization of one can be rotated without affecting the other. Hence the intermediate N layer which of course should be a "vegetable", neither superconducting nor magnetic, is required. One can visualize the structure, which is schematized in Fig. 2.1, by considering Fig. 5.1 and mentally inserting a normal layer between the two ferromagnets.

As to the materials used, the F layers were made [21] of cobalt, the superconductor was niobium and the intermediate N layer was copper. These were chosen largely because of the experimenters' ability to make uniform layers with clean interfaces using these specific materials.

6.1 Basic Experimental Description

Despite the theoretical emphasis of this book, it is relevant to the comparison with experiment to give here some additional experimental information about the samples: the interested reader can see more details in Ref. [21] The basic structure of the samples discussed in this chapter is displayed in Fig. 6.1. This figure, and all the others in this chapter, is extracted from Ref. [21]. The layers are shown in an exploded view. The thicknesses of the two Co layers are denoted as d_p and d_f. These indices stand for "pinned" and "free" respectively: the internal field of the outer layer is pinned by the presence of the CoO top layer which, at low temperatures,

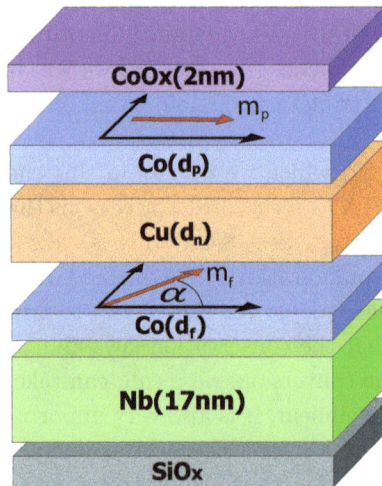

Fig. 6.1 Schematic of the experimental samples discussed in this chapter. The relevant part of the sample consists of the Co, Cu, Co and Nb layers. A silicon oxide substrate is also shown and a cobalt oxide layer covers the top of the sample. The thicknesses of the relevant layers are indicated.

is antiferromagnetic. The magnetization of the "free" layer can then be rotated by the indicated angle α via the application of a small magnetic field. The angle α could be and was varied between zero and 180 degrees in a reproducible way by applying a small magnetic field.

The thickness of the Cu and both of the Co layers could be carefully controlled when growing the various samples. The Nb thickness was fixed in all cases to 17 nm, a value comfortably higher (as discussed more fully below) than the effective coherence length. Several series of samples were manufactured. In one, d_p was fixed to 2.5 nm, d_n to 6 nm and d_f was varied between 0.5 and one nm at approximately 0.1 nm intervals. In a second series, d_n was varied between 4 and 6.8 nm with d_f and d_p fixed at 2.5 and 0.6 nm respectively. The third series of samples investigated the d_f dependence in the range from 1.5 to 5.5 nm, with fixed d_p and d_n being 0.6 and 6 nm respectively. So each series focused on the dependence of one of the three relevant lengths, while in all cases the α dependence was also investigated.

This is not the right venue, nor do I have the required expertise, to discuss the sample growth details. I can say however that the samples were of the best quality. It is of course impossible to make samples with *perfect* interfaces, but these are very good indeed, as we shall see from the parameter discussion below: interfacial scattering is minimized. Note that even with a hypothetical geometrically perfect interface, interfacial scattering due to Fermi wavevector mismatch will occur, as can be inferred from elementary quantum mechanics considerations. The interested reader should consult Ref. [21] for details of the growth procedure, sample characterization, and the techniques used to measure the transition temperature.

6.2 Some Theoretical Details

The methods for the calculation of the transition temperature have been discussed in detail in Chap. 4, with detailed examples given in Secs. 4.2.1 and 5.1. The non-monotonicity of T_c with respect to the misalignment angle has been seen in Fig. 5.2 and discussed in Sec. 5.1 for the simplified spin valve structure. Those examples pertain to simple trilayer structures, while here we have to deal with a more complicated five layer system, including the presence of two ferromagnetic layers and one normal layer. In the geometry of the experimental system studied the BdG equations (see Chap. 2) are of the form,

$$\begin{pmatrix} \mathcal{H}_0 - h_z(y) & -h_x(y) & 0 & \Delta(y) \\ -h_x(y) & \mathcal{H}_0 + h_z(y) & \Delta(y) & 0 \\ 0 & \Delta(y) & -(\mathcal{H}_0 - h_z(y)) & -h_x(y) \\ \Delta(y) & 0 & -h_x(y) & -(\mathcal{H}_0 + h_z(y)) \end{pmatrix}$$

$$\times \begin{pmatrix} u_{n\uparrow}(y) \\ u_{n\downarrow}(y) \\ v_{n\uparrow}(y) \\ v_{n\downarrow}(y) \end{pmatrix} = \epsilon_n \begin{pmatrix} u_{n\uparrow}(y) \\ u_{n\downarrow}(y) \\ v_{n\uparrow}(y) \\ v_{n\downarrow}(y) \end{pmatrix}, \tag{6.1}$$

which is of the generic Eq. (2.8) form. The notation $h_i(y)$ ($i = x, z$) denotes the components of the exchange fields $\mathbf{h}(y)$. We have $h_x(y) = h_0 \sin(-\alpha/2)$ and $h_z(y) = h_0 \cos(-\alpha/2)$ in F_f, the free layer. In the pinned layer, F_p, we have $h_x(y) = h_0 \sin(\alpha/2)$ and $h_z(y) = h_0 \cos(\alpha/2)$. Here h_0 is the magnitude of the exchange field, assumed of course to be the same in the two Co layers. See Table 2.1 for normalization details.

In Eq. (6.1), the single-particle Hamiltonian (see Eq. (2.9)) contains an effective interfacial scattering potential described by delta functions of strength H_j (j denotes the different interfaces of which there are now three), so that one has (cf. Fig. 6.1):

$$U(y) = H_1\delta(y - d_S) + H_2\delta(y - d_S - d_f)$$
$$+ H_3\delta(y - d_S - d_f - d_n), \tag{6.2}$$

where $H_j = k_F H_{Bj}/m$ is written in terms of the dimensionless (see Sec. 2.6 and Table 2.1)) scattering strengths H_{Bj}. As in Chap. 2 Eq. 6.1 must be solved iteratively including the self-consistency condition, Eq. (2.12), which formally remains the same. The matrix elements must of course be recalculated.

In defining the internal magnetic fields and the angle α above we have assumed that the quantization axis lies along a specified z direction, but one can easily obtain the spin-dependent quasiparticle amplitudes with respect to a different spin quantization axis rotated by an angle θ in the $x - z$ plane via the spin rotation Eq. (3.8) matrix:

$$\hat{U}_0(\theta) = \cos(\theta/2)\hat{\mathrm{I}} \otimes \hat{\mathrm{I}} - i\sin(\theta/2)\rho_z \otimes \sigma_z, \tag{6.3}$$

where ρ and σ are again vectors of Pauli matrices in particle-hole and spin space respectively.

6.3 Results

6.3.1 *Transition temperature*

The theoretical value of T_c is then determined from the eigenvalue procedure described in Sec. 4.2.1. It is not hard to see that the only significant difference between the calculation of T_c in the present case and the example given in Chap. 4 is in the (admittedly gory) details of the calculation of the matrix elements, denoted as J_{lk} in Eq. (4.4). The explicit expression for these matrix elements as obtained from Eq. (4.5), which remains formally valid, is different from that one would have for the SFS or FFS systems.

One must judiciously choose the material parameter values appropriate to the case. This is a rather complicated process. Some parameters are well known from experimental characterization, others are reasonably known, at least within a range, while others have to be varied, within reasonable limits, to fit the appropriate values. For every parameter set, one must evaluate T_c numerically as a function of the misalignment angle α. This makes any kind of least squares fit unfeasible. Therefore, one searches within plausible regions of parameter space, and finds a fit that is the best among the sets tried but not necessarily (unless one be very lucky) the *best* possible fit. There are a number of parameters at one's disposal when computing the theoretical values of T_c, as one can see by considering Table 2.1 and extending it mentally to the preset five layer case. One first has to strive to keep the number of fitting parameters as small as possible. All of the relevant physical parameters that are related to the properties of the materials involved, such as the exchange field, and the effective superconducting coherence length, are required to be the same for all of the different samples when performing the fitting. However, for parameters that are affected by the fabrication processes, such as the interfacial barrier strength, one can reasonably assume that their values are somewhat different from sample to sample. One of course finds that the variation is small between different samples in each series since after all they were all fabricated in the same way. For the material parameters it was found [21] that the best value of the effective Fermi wave vector was $k_F = 1$ Å^{-1} and the effective superconducting coherence length $\xi_0 = 11.5$ nm. For the dimensionless exchange field $h \equiv h_0/E_F$ (normalized to the Fermi energy), the effective value for Co is taken to be $h = 0.145$ which is consistent with the work reported in Chap. 5. For the superconducting transition temperature of a putative pure superconducting sample of the same quality as the Nb material in the layers, the value $T_c^0 = 4.5$ K had been found [22] in previous experimental work by the same group. It is of course lower than the bulk transition temperature of pure Nb. All of these parameters must be and are kept invariant across all of the different samples, as mentioned above. Only the three interfacial barrier strengths can be treated as somewhat adjustable from sample to sample during the fitting process. It is assumed, however, that the barrier strength is the same on both sides of the normal metal layer, while that between the free ferromagnetic layer and the superconductor is weaker. For each series, the barrier was indeed found to vary somewhat as the relevant thickness changed. The best values reported for the dimensionless barrier parameters, as used here, are as follows: $H_{B1} = 0.2$, and both H_{B2} and H_{B3} vary from 0.64 to 0.7 for different batches in the d_f series. For the d_p series, one has $H_{B1} = 0.15$, $0.53 < H_{B2}$, and $H_{B3} < 0.58$. The d_n series samples were best fit at H_{B1} ranges from 0.3 to 0.45 and $H_{B1} = H_{B2} = 0.62$. The thicknesses of the different layers are taken of course from their experimental values. As is usually the case in such structures, a thin magnetic "dead layer" was found to exist between the normal metal and the free ferromagnetic layer; its thickness was in the range $0.27\,\text{nm} - 0.35\,\text{nm}$.

We now discuss the comparison between the experimental and theoretical values of T_c as a function of layer thicknesses and angle α for the three different batches of samples: recall that in the first one varies d_f, in the second, d_p and in the last, d_n. First, in Fig. 6.2, we see comparisons between experiment and theory, for the T_c results in the parallel state ($\alpha = 0$), as a function of the varying thickness for the three different series as just mentioned. In all three series, the experimental and theoretical T_c values are in

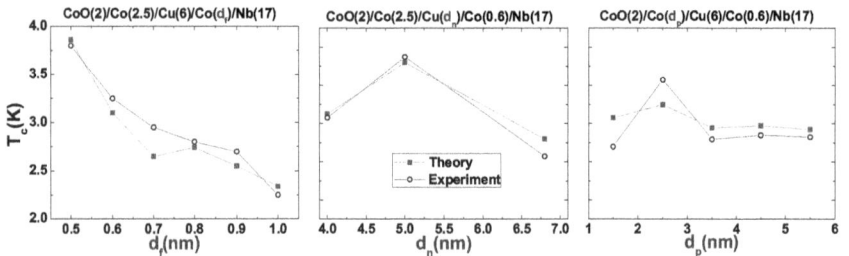

Fig. 6.2 Experiment and theory comparisons of T_c in the parallel state ($\alpha = 0$) are shown for the three batches of samples: Left panel: six different Co free layer thicknesses as shown, with $d_p = 2.5$ nm, $d_n = 6$ nm. Middle panel: three different Cu layer thicknesses, with $d_f = 0.6$ nm and $d_p = 2.5$ nm. Right panel: five different Co pinned layer thicknesses, with $d_f = 0.6$ nm, $d_n = 6$ nm.

very good agreement with each other. For the d_f series, one should notice that both experimental and theoretical T_c results are very sensitive to the thickness of the free layers. When the thickness of the free ferromagnetic layer is increased, T_c decreases non-monotonically by almost 50%. However, the d_n and d_p series do not show the same sensitivity, even though the ranges of thicknesses for these two series are larger compared to that of the d_f series. This lower sensitivity is physically reasonable: because of the presence of ferromagnets, the magnitude of the singlet pairing amplitude in non-S regions away from the F/S interface decreases very fast beyond the boundary.

The exchange field reduces the proximity effect. Therefore, the size effects from the thicknesses of normal metal layers and fixed ferromagnetic layers are less prominent. One should again emphasize that both theoretical and experimental T_c are usually found to depend non-monotonically on the thicknesses of the F layers. In fact, except for the T_c for d_f series, which does not show any strong oscillatory behavior, all other series exhibit the non-monotonicity of T_c. Oscillatory behavior of transition temperatures as one varies the thickness is standard in hybrid S/F heterostructures due to the oscillatory character of the pair amplitude as pointed out in Chap. 4 and earlier. The reason for the exception found might be that the data points are too widely spaced.

Next, in Fig. 6.3, there is a detailed comparison of theoretical and experimental results for $\Delta T_c(\alpha) \equiv T_c(\alpha) - T_c(0)$ as a function of the angle α between the magnetizations in the free and fixed layers. Results are given for the d_f, d_n, and d_p series. Each panel in the first row in Fig. 6.3 represents different samples for the d_f series, as labeled. Results for the d_n and d_p series are plotted in the second and third row respectively. One can clearly see that the behavior of the highly non-monotonic angular dependencies of the theoretical results describe very well the experimental results, not only qualitatively but also quantitatively. The magnitudes of the experimental and theoretical results for ΔT_c are comparable: both experimental and theoretical results indicate that the switching effects are in about 25 mK range. For the d_f series, we see that the switching range for both experimental and theoretical $T_c(\alpha)$ varies non-monotonically when d_f is increased. This occurs for the same reason already mentioned in the discussion of Fig. 6.2: the behavior of $T_c(\alpha)$ is very sensitive to the inner ferromagnetic layer thicknesses due to the proximity effects. Similarly, the switching ranges are less sensitive to the thickness of the outer ferromagnetic layer both in the d_p series and also to the normal metal layer thickness in d_n series.

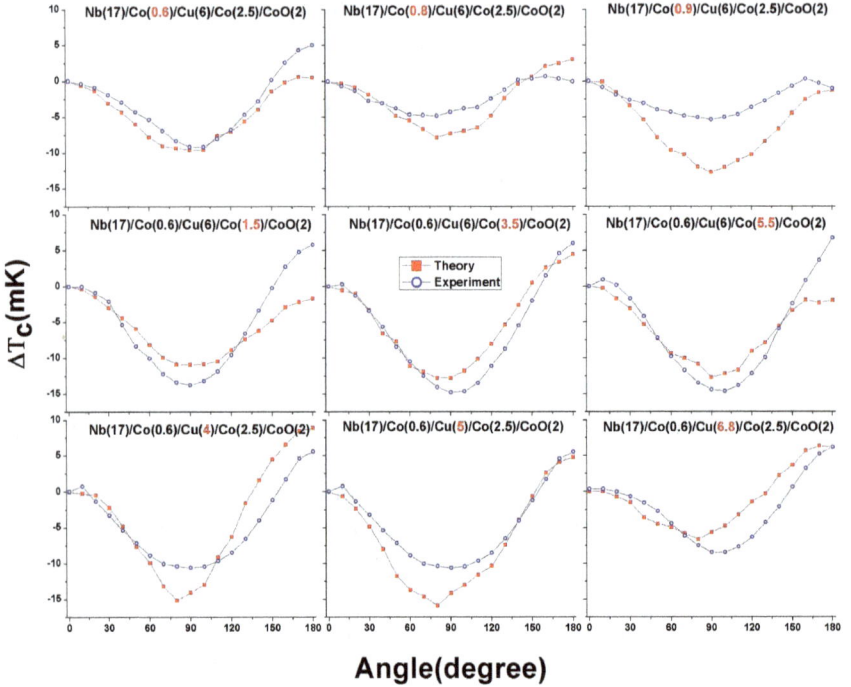

Fig. 6.3 Experiment and theory comparisons of ΔT_c [defined as $\Delta T_c(\alpha) \equiv T_c(\alpha) - T_c(0)$] as a function of relative magnetization angle α are shown for the three batches of samples. Top row: three different free layer thicknesses, $d_f = 0.6$ nm, 0.8 nm, 0.9 nm, with $d_p = 2.5$ nm, $d_n = 6$ nm. Middle row: three different pinned layer thicknesses, $d_p = 1.5$ nm, 3.5 nm, 5.5 nm, with $d_f = 0.6$ nm, $d_n = 6$ nm. Bottom row: three different nonmagnetic layer thicknesses, $d_n = 4$ nm, 5 nm, 6.8 nm, with $d_f = 0.6$ nm, $d_p = 2.5$ nm.

6.3.2 *Triplet conversion*

We now turn to a very important discussion of the correlation between the behavior of the superconducting transition temperatures and the existence of odd triplet superconducting correlations in these systems. From the self-consistent wavefunctions and eigenvalues one can readily compute the induced triplet pairing amplitudes which we denote as $\mathbf{f_0}$ (with $m = 0$ spin projection) and $\mathbf{f_1}$ (with $m = \pm 1$) according to the expressions Eq. (3.1) given in Chap. 3. As mentioned there, these triplet pair amplitudes are odd in time t and hence vanish at $t = 0$, in accordance with the Pauli exclusion principle.

The role that induced triplet correlations play in the nonmonotonic behavior of $T_c(\alpha)$ is fundamental. One can see that by comparing theoretical results for the triplet amplitudes and experimental results for the transition temperature. To examine this question in a quantitative way, one first must calculate the induced odd triplet pairing correlations for all of the thickness values and angles studied in the experiments. These correlations (and of course the ordinary singlet correlations as well) are self-consistently calculated using the methods previously described. As noted in Chap. 3, with the presence of non-homogeneous magnetizations the triplet pair amplitudes in general can be induced when $t \neq 0$. It is convenient to perform the study of the overall effect in terms of the quantity

$$F_t(y,t) \equiv \sqrt{|f_0(y,t)|^2 + |f_1(y,t)|^2},\qquad(6.4)$$

where the quantities involved are defined in Eq. (3.1). This quantity accounts for both triplet components: the equal spin and opposite spin triplet correlations. The reason to use this quantity is that via Eq. (6.3), one can easily show that, when the spin quantization axis is rotated by an angle θ, the rotated triplet pair amplitudes $\tilde{f}_0$ and $\tilde{f}_1$ after the transformation are related from the original f_0 and f_1 by:

$$\tilde{f}_0(y,t) = \cos(\theta)f_0(y,t) - \sin(\theta)f_1(y,t),\qquad(6.5a)$$
$$\tilde{f}_1(y,t) = \sin(\theta)f_0(y,t) + \cos(\theta)f_1(y,t).\qquad(6.5b)$$

Therefore, using the quantity $F_t(y,t)$ obviates any ambiguity issues related to the existence of generally non-collinear "natural" axes of quantization in the system.

This quantity $F_t(y,t)$ can be computed as a function of position and of α. It turns out to be particularly useful to focus on the spatially averaged value of $F_t(y,t)$ in the pinned (outer) layer F_p. This is the portion of the sample farthest away from the superconductor, and the triplets there are the ones that, so to speak, have traveled farthest from their source. It is natural to normalize this averaged quantity, computed in the low T limit, to the value of the singlet pair amplitude in the bulk S. This normalized averaged quantity is plotted, as a function of α in Fig. 6.4 (left vertical scale, see caption) at a dimensionless characteristic time $\tau = 4.0$. This time value is unimportant, provided it be nonzero, of course.

In each of the three panels in Fig. 6.4, an example is taken from each of the three experimental series discussed in Sec. 6.1, as explained in the

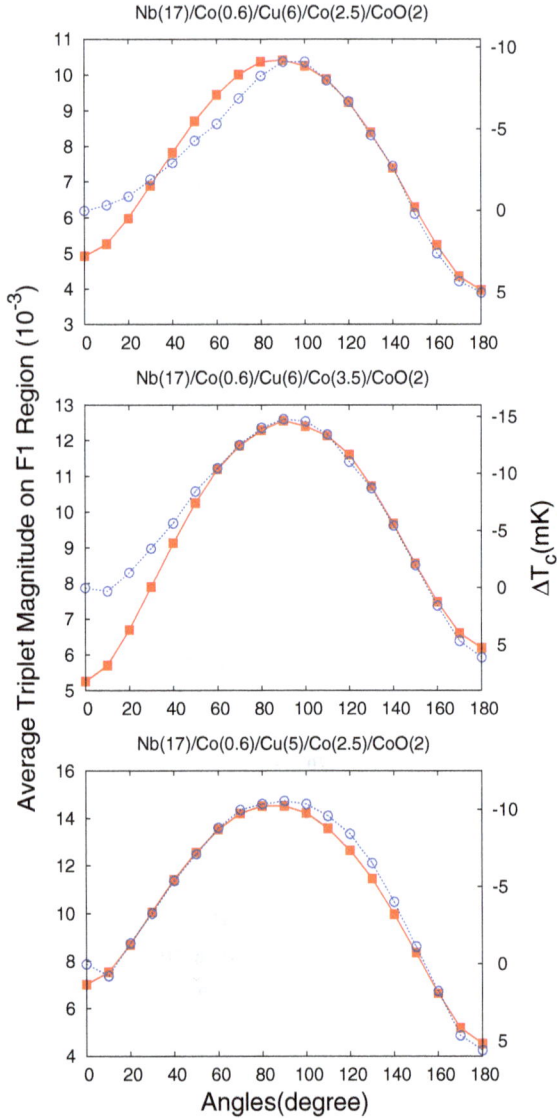

Fig. 6.4 Two correlated quantities both plotted as a function of α. Left vertical scale and square red, solid data points: theoretically calculated average triplet amplitudes in the pinned ferromagnet layer in the spin valve samples described. The quantity plotted is the average of $F_t(y, t)$ (Eq. (6.4)) in this pinned ferromagnetic region. Right vertical scale (note that it is inverted) and blue, empty circles: the experimental ΔT_c. The ΔT_c data corresponds to one set chosen from each batch of samples (see text). Top panel: from the d_f series, Middle panel: from the d_p series, Lower panel: from the d_n series.

caption. One can observe that the maxima of this average F_t occur when $\alpha = \pi/2$ and its minima are at either $\alpha = 0$ or $\alpha = \pi$. This is because there are no triplets with $m = \pm 1$ for the parallel or antiparallel config-urations. The curve is not symmetric around $\pi/2$ because the zero and π configurations are not the same. Now comes the fundamental point: in the same figure (right vertical scale) the experimental values of $\Delta T_c(\alpha)$, for the same cases, are plotted in an inverted scale. These plots also have minima near $\pi/2$, which means that T_c is maximum there. Since the units for the two quantities plotted are different, the vertical axes were adjusted by slid-ing one of them so that the points near $\alpha = \pi/2$ coincide. The agreement in the shape of the curves is truly striking and shows quantitatively how T_c is anti-correlated with the triplet density. This anti-correlation can be easily understood: the magnitude of the low T singlet pair amplitudes is of course positively correlated to T_c. That triplet pair amplitudes are anti-correlated to T_c (or to the singlet amplitudes) indicates a singlet-triplet conversion process: when more singlet superconductivity leaks into the fer-romagnet side, T_c (which is of course determined by what happens in the singlet channel of Nb) is suppressed and triplet amplitudes are enhanced. The average magnitude of the triplet pair amplitudes in the free and normal layer regions is only weakly dependent on α: what is important is the long range propagation of triplet pairs throughout the entire system, generated by the symmetry-breaking interfaces and magnetic inhomogeneity created from the two misaligned ferromagnets. This clearly demonstrates a singlet to triplet process which is related to the non-monotonicity of the transition temperature in a **quantitative** way.

6.4 Summary

This has been a very important chapter. In it, we have first verified that this book's methods can be used to make very detailed quantitative pre-dictions for the transition temperature of multilayer spin-valve structures as a function of the geometry (thickness of the layers) and on the misalign-ment angle between the magnetization directions of the two ferromagnets involved. These predictions are in excellent agreement with very extensive and detailed experimental results.

Furthermore, in Sec. 6.3.2 we have seen that the presence of triplet pairs *quantitatively* accounts for the variation of the transition temperature with angle in these systems. The importance of this quantitative understanding is that it is a direct proof that triplets exist. Other experimental proofs of the existence of triplets are based in the existence of proximity effects which are long-ranged, such as tunneling that occurs over distances that are too long to be plausible for singlet pair tunneling. Such experiments are always open to criticisms of the "your sample has a pinhole" variety. Measurements of T_c are quite immune to such criticisms.

This chapter concludes also the first part of the book, dealing with thermodynamics. We will next consider transport properties.

Chapter 7

Transport: Introduction

7.1 Basic Ideas

Until now, we have studied in this book the thermodynamic properties of Ferromagnet/Superconductor systems. The structures we have considered carry no current: the wave functions that we have considered are real and, furthermore, they vanish at the sample surface. This is all well and good to study a sample disconnected from the rest of the world, but not to consider transport. And most of the interesting potential practical applications of these systems involve connecting them to things and driving currents through them.

This chapter is largely a pedagogical outline of how one deals with transport in S/F structures. Its purpose is to make the subsequent developments easier to understand. Many details will be left somewhat vague. These details will be discussed and vagueness dispelled in subsequent chapters, chiefly Chap. 8 for charge transport and Chap. 9 for spin.

To make a road map of what we have to do now, consider elementary Quantum Mechanics in one dimension. Suppose we have a particle with a certain potential, such as a barrier. Assume, for a moment, that the particle is confined to a certain large but finite region. In effect, this would amount to assuming that at the edges of the region studied rise the walls of an infinite potential well. The static properties (the spectrum) of the system could be studied, for any potential, via Fourier analysis expanding in a complete set of functions which vanish at the edges of the region studied. This is what is accomplished, in our case, by the choice of sine functions in Eq. (2.14): the choice implies that electrons and holes are confined to a region which is finite in the direction normal to the layers.

Next, return to the simple quantum mechanical example but imagine now that the particle is not confined: it enters from (say) the left, scatters from the potential barrier and is then partly transmitted and partly reflected. To study this case we would have to consider an incoming plane wave plus a reflected wave at the far left plus a transmitted wave at the far right. Not only we would have to consider plane waves, instead of sine functions, but we would have to adjust boundary conditions so that e.g., the wavefunction on the right is purely transmitted. In undergraduate quantum mechanics we learn how to do this "by hand" by looking for solutions of the required form: a normalized incoming plane wave plus a reflected wave (amplitude R) at the left, and a transmitted wave, of amplitude T at the right. To do so for complicated scattering potentials (as opposed to the very simple examples in textbooks) is a little messy but quite doable. Typically, the quantity wanted is the transmission coefficient $\mathcal{T}$. This is the square of the amplitude T of the transmitted wave. Using particle conservation one has that

$$\mathcal{T} = 1 - |R|^2. \qquad (7.1)$$

In our case, the situation is much more complicated. We have of course a large number of quasiparticles in the problem, and although they are assumed to behave rather independently this is a very substantial complication. Our system is three dimensional. Treating it, as we do, in the limit where it is very large in the transverse directions, (the quasi one-dimensional limit) eliminates only some but by no means all of the complications associated with the three-dimensionality. Furthermore, there are both particles and holes and, via the pair potential term in the Hamiltonian, particles are changed into holes and vice-versa. Then of course there is spin. Even in the simpler case, where scattering at the interfaces is not spin-dependent, one still has to deal with the existence of singlet to triplet conversion. And we will want to deal with spin transport in addition to ordinary charge currents.

One thing that remains the same, however, is that we must at the end deal with appropriate transmitted and reflected currents, and for the case of charge transport (spin will prove more complicated) with some kind of transmission coefficient. In fact, the relevant quantity for charge transport is, as we will see, the *conductance*, that determines the probability of charges (not quasiparticles) getting through the system. It is conductance, not conductivity, since it is clear here that size matters: we are talking about an extensive property.

7.1.1 *Reflection, Andreev scattering and boundary conditions*

As just mentioned, in the simple quantum mechanical example one considers a particle coming from the left, and normalizes the corresponding current to unity. Then, there are two possibilities for the particle as it scatters from some potential: either it gets transmitted or it gets reflected. Each of these processes has some amplitude and the absolute value squared of the transmission amplitude is basically what we need.

Next, as an additional pedagogical exercise, let us consider a system consisting of a semi infinite ferromagnet (occupying, say the region $y < 0$) and a superconductor similarly occupying the $y > 0$ region. If one throws away self-consistency and makes the assumption that the pair amplitude is a constant in S and zero in F then one can solve the problem analytically, with a bit of patience and care. Note however that the step function assumption as to the pair potential (apart from throwing away the proximity effect) violates self-consistency, which as we shall see below in Sec. 7.2, can lead to serious violations of conservation laws, in general a very dangerous thing when one tries to calculate transport properties.

But to obtain such a solution one must consider all of the possible scattering outcomes and now there are many of them. Suppose one has an incoming particle with spin up and a certain energy ϵ and a corresponding wavevector. Given the spin of the particle and the value of ϵ, this wavevector is known. This particle may of course undergo the ordinary processes of being transmitted, as a particle with the same spin, or being reflected, again as a particle of the same spin. But more can happen. It can be reflected as a hole, with spin opposite to what the incoming particle had: this is the famous Andreev reflection process, as shown in Fig. 7.1. This process is also called SaintJames–Andreev reflection. Both authors published their work at about [23, 24] the same time, but SaintJames did it in a French journal, so he got less exposure. It also can be transmitted as a hole-like excitation of opposite spin. So, there are four processes all together, plus four more for the opposite spin. For a given wave vector and energy of the incoming particle one can work out with a little patience, the wavevectors corresponding to the four processes listed, (see Eq. (8.3) if you want to peek ahead) remembering to properly take into account that when the spin flips the particle in the ferromagnet goes to the opposite spin band. There are many interesting and amusing things one can discover by digging further: for example, if the incident wavevector forms an angle θ with the normal to the interface, then the ordinarily reflected particle does have a wavevector

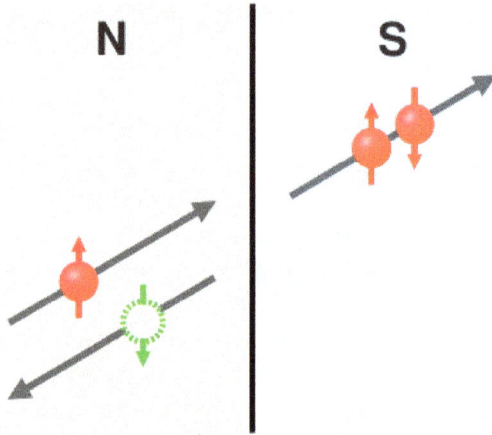

Fig. 7.1 The famous Andreev (or Andreev–SaintJames) reflection process at a Normal/ Superconductor interface: the incoming particle (solid) with spin up is reflected as a spin down hole, (empty circle), while a Cooper pair forms in the S side. This is an approximate description.

pointing towards $-\theta$, but the direction of the wavevector of the Andreev reflected hole in F is different. One can discuss then things such as angles of total reflection, as in elementary optics. There is no point spending time here on these points, since the approximation that they involve is geometrically (infinitely thick slabs) and physically unrealistic. The important thing to learn is that there are many processes possible, that all must be included properly and, most of all, that to evaluate the correct transmission coefficient (that is, to evaluate the conductance) one needs to consider boundary conditions corresponding to a properly normalized incoming particle with definite spin and wavevector.

Now, let us call the amplitude of the wave reflected by ordinary reflection b_σ, where we have added a spin index because in general it will depend on spin. Why? Because particles of the same energy in the up and down spin bands in the ferromagnet will have different wavevectors and we do know, again from elementary quantum mechanics, that reflection and transmission coefficients depend on wavevector. As we have seen, there is also an Andreev reflected wave, let us call its amplitude a_σ. One might too hastily conclude that the generalization of Eq. (7.1) is simply $1 - |b_\sigma|^2 - |a_\sigma|^2$. But this would be wrong, because the Andreev reflected entity is a hole, and has opposite charge: we are looking for *charge* transmission, not *number* transmission. So the sign of the Andreev contribution, we see, should be reversed and we should write $1 - |b_\sigma|^2 + |a_\sigma|^2$. But this would still not be right: the transmission and reflection coefficients are ratios of the transmitted and reflected *currents* to the incoming current, and the currents have an extra wavevector factor. In the elementary quantum mechanics example of a localized barrier, all wavevectors are the same and they cancel. This is of course not the case here, just as it is not the case in elementary quantum mechanics for scattering from a potential step: so one has to multiply by the ratio of reflected (Andreev or ordinary) to incoming wavevectors. Hence we conclude that the contribution from these scattering processes to the charge transmission is of the form:

$$\mathcal{G}_\sigma(\epsilon) = 1 - r_b |b_\sigma|^2 + r_a |a_\sigma|^2 \tag{7.2}$$

where the quantities r_b and r_a are the ratios of the wavevectors corresponding to Andreev and ordinary reflections to that of the incoming one. These ratios and their spin dependence can be sorted out with a bit of patience and we will get into the details of how one does it in the next chapter. Of course one has to sum the result of Eq. (7.2) over spins, with appropriate weights, and over wavevectors, for a given energy. We will also get into this later.

Equation (7.2) is a version of the famous BTK (Blonder, Tinkham, and Klapwijk) [25] formula. The original reference has a more rigorous, but much lengthier, derivation: do consult it if you are not satisfied with the intuitive derivation just given. Note also that the derivation given here being a purely single-state quantum mechanical argument, it is the zero temperature result. We will return later to the question of the temperature dependence, which is nearly always neglected in the literature, not always for sufficiently good reason.

The point of this exercise is that for the infinite thickness, non-self-consistent problem introduced above, the amplitudes can be found analytically. One simply writes the appropriate wavefunctions. These are four-component vectors (the components corresponding to particle/hole and spin up/down). One then determines the coefficients by demanding continuity of the wavefunctions at the interface, $(y = 0)$ plus enforcing a discontinuity in the derivative related to a delta function potential arising from interfacial scattering. This is rather tiresome but not strictly difficult. It need not be done here since we will, in the next chapter, present the method to solve our more complicated finite thicknesses problem, correctly and self-consistently.

One should ask next: what does Eq. (7.2) represent as a contribution to the conductance, which is what one measures experimentally? The answer is simple: each transmitted charge contributes one conductance quantum G_q to the result. This will be, therefore, the proper way to normalize conductance, as anticipated in Table 2.1. Hence, the result of summing results from Eq. (7.2) is the conductance itself in units of the conductance quantum per spin, (e^2/h) or $3.87 \ 10^{-5}$ S. This follows immediately from the textbook Landauer formula.

In the above arguments we have used the boundary condition that, from the left, we have incoming particles and also ordinarily reflected and Andreev reflected quasiparticles and holes. At the right end, in the superconductor, we would have particles moving to the right (with amplitudes given by u-type coefficients) and holes to the left (with v-type coefficients). Each of these terms would of course be multiplied by some transmission amplitude, The overall transmission could evidently be written in terms of the different transmission amplitudes, instead of the reflection terms. We have done it here in terms of reflected waves because it is a little easier, since then the coefficients corresponding to incoming waves are unity (because of the chosen normalization) and need not be found. We will see the precise details in Chapter 8.

7.1.2   *The voltage dependence of the conductance*

Continuing now with the pedagogical heuristic arguments which will later be justified by detailed calculations, let us have here a qualitative discussion of how we expect the conductance to behave. As mentioned just below Eq. (7.2) one sums the conductance over spins and wavevectors at constant energy. The current is driven by a voltage V so the energy of the incoming particles will be eV measured from the Fermi surface. We will see in a moment that the relevant energy scale is the gap in S, which at this point is assumed to be Δ_0, but eventually should be computed self-consistently. Therefore it is convenient to define a dimensionless voltage as $E \equiv eV/\Delta_0$. This is an exception to our policy of normalizing the energies to the bandwidth E_F but it is wise to make this one exception, as the physics demands it. So, we wish to see what the conductance function $G(E)$ may look like.

The first question that might come to the mind is: why should G depend on voltage at all? For a chunk of metal, and for an applied voltage which is not too large, there is no such dependence: this is Ohm's law, which says that current is proportional to voltage, the coefficient of proportionality being independent of V. But Ohm's law is an empirical result; it does not follow from Maxwell's equations or from any fundamental law of nature. So, let's think a little:

If the voltage is much larger that Δ_0/e, $E \gg 1$, then clearly the pair potential cannot make any difference and $G(E)$ will tend to a constant that will be of order unity (i.e., one quantum per channel with the conventions discussed above) if interfacial scattering is negligible, or a much smaller value if the scattering is large, the tunneling limit. How about the limit $E \ll 1$? Well, then it depends. If the interfacial scattering is small, then the proximity effects will be very prominent, Andreev scattering will be strong and we would expect a normalized value per channel of order two. Two and not one because of the extra Andreev retro-reflected hole. We can then imagine that when the applied voltage is increased, nothing much would happen until E approaches unity, at which point the conductance would drop to unity, probably not abruptly but rather smoothly. This is called the Andreev limit. On the other hand, if interfacial scattering is extremely strong, then there would be basically no proximity effect; G would be exponentially small at small E. This is called the tunneling limit (annoyingly enough, many people call the conductance of these layered systems the "tunneling conductance" even in the Andreev limit). It can be understood in a model where one treats the superconductor as if it were a semiconductor with a gap. $G(E)$ then increases with E until E reaches unity, where it has a peak (due to the density of states singularity) and drops to the (small) large voltage value. In reading the literature one should keep in mind that because in the old days samples with good quality interfaces were not available, many older elementary textbooks [3, 26] discussed only the tunneling limit, ignoring the opposite Andreev situation.

Now, again qualitatively, what do we expect will happen in realistic samples, as considered in this book? Recall that the samples we are interested in are good quality samples, such as those discussed in Chap. 6. Good does not mean perfect: some interfacial scattering due to minor imperfections and to wavevector mismatch is unavoidable. Also, because of the proximity effect, the pair potential does not change abruptly at the interface. There, it has a value considerably smaller than the bulk value of Δ_0. Hence, there cannot be any abrupt changes at $E = 1$ but only, perhaps, a region of relatively rapid changes near that value.

Therefore, at zero bias G can be expected to have some value that may be larger than unity, but not reaching two, or smaller than unity, but not exponentially small. This is called the zero bias conductance value,

ZBC (acronyms are collected in Table 8.1). Then, in the region $0 < E < 1$ (conventionally called the "subgap" region), $G(E)$ will go up or down, having a peak, perhaps at some intermediate value. We will see that in fact in some cases it may have more than one peak. The value $G(E = 1)$ is not special because of the proximity effects, but beyond that G will fall, or rise, to the normal metal value. The exact results then depend on geometry and all kinds of other details, as we shall see. But it is good to keep in mind these qualitative expectations.

7.2 Conservation Laws

We discuss now the important issue of the charge conservation laws. In transport calculations, we all know from textbooks that it is fundamental to assure that they are not violated [27] as otherwise major disasters will ensue: one cannot assume that conservation laws are violated "just a little".

Let us begin from the Heisenberg equation

$$\frac{\partial}{\partial t} \langle \rho(\mathbf{r}) \rangle = i \langle [\mathcal{H}_{e\!f\!f}, \rho(\mathbf{r})] \rangle, \tag{7.3}$$

where the angular brackets denote the proper averaging. The above commutator can be computed without difficulty since the operators involved are fermionic. One arrives at the following continuity condition:

$$\frac{\partial}{\partial t} \langle \rho(\mathbf{r}) \rangle + \nabla \cdot \mathbf{j} = -4e\text{Im} \left[\Delta(\mathbf{r}) \left\langle \psi_\uparrow^\dagger(\mathbf{r}) \psi_\downarrow^\dagger(\mathbf{r}) \right\rangle \right]. \tag{7.4}$$

In the steady state, which is what we are concerned with here, the first term on the left is zero. Equation (7.4) is then simply an expression for the divergence of the current. One can proceed now in a quite general way, but it is simpler and equally enlightening to specialize to a quasi one-dimensional system. Then, the current has only one component, j_y, which depends on y only. We have again taken the y-axis normal to the layers. Expressing the ψ operators in terms of the wavefunctions the conservation law can be rewritten as:

$$\frac{\partial j_y(y)}{\partial y} = 2e\text{Im} \left\{ \Delta(y) \sum_n \left[u_{n\uparrow}^* v_{n\downarrow} + u_{n\downarrow}^* v_{n\uparrow} \right] \tanh \left(\frac{\epsilon_n}{2T} \right) \right\}. \tag{7.5}$$

Charge conservation implies that the right hand side of this equation ought to be zero. When the system is in equilibrium the self-consistency condition, Eq. (2.12) in Sec. 2.5, on the pair potential indeed causes the right hand side of Eqs. (7.4) or (7.5) to be zero. This would **not** necessarily be the case if a non-self-consistent solution were used. It is true that the quantity involved vanishes for real wavefunctions, but one must recall immediately that given the boundary conditions involving all kinds of incoming, transmitted and reflected plane waves one does not have real wavefunctions in the transport case.

This relation between self-consistency and charge conservation can be readily derived as a simple exercise for bulk BCS superconductors and has been known since the early nineteen sixties but, curiously, it keeps being forgotten. Calculations involving non-self-consistent pair potentials keep popping up in the literature to this day in many contexts. The calculations thus performed are much simpler than the correct ones but the attentive and careful reader should view all such results as garbage. Indeed, even in simple $N - S$ heterostructures, self-consistency is crucial to account properly for all of the Andreev scattering channels arising when the current is constant throughout the system. This makes sense: Andreev refection accounts for a large part of the proximity effect.

One can also write explicit expressions for the charge density ρ and the current $j(y)$. In the problem we are considering, there exists a finite voltage bias V between the two ends of the system. This finite bias leads to a non-equilibrium quasi-particle distribution and results of course in a net current. Still, charge conservation must hold. To see how this works in this non-equilibrium case we first write down the net quasi-particle charge density in the $T \to 0$ limit (the case we consider for now, extension to $T > 0$ is a simple matter of adding some Fermi function occupation factors) by considering the excited state $|\mathbf{k_1 k_2} \cdots \rangle$ caused by the energy shift eV due to the finite bias V. Thus, this excited state contains all single particle states $|\mathbf{k_j}\rangle$ $(j = 1, 2, \cdots)$ with energies less than eV. Pedagogically, let us first consider the contribution arising from a single-particle state. Let us denote by $|\mathbf{k}\rangle$ this single particle state with an incident wavevector $\mathbf{k} = \mathbf{k_\perp} + k\hat{\mathbf{y}}$ and energy $\epsilon_{\mathbf{k}}$. The charge density associated with it is written as

$$\rho = -e \sum_\sigma \left(|u_{\mathbf{k}\sigma}|^2 - |v_{\mathbf{k}\sigma}|^2 \right). \tag{7.6}$$

Physically, the first term represents the ground state charge density. For a generic excited state, $|\mathbf{k_1 k_2} \cdots \rangle$, that can contain many single-particle states, one need to sum over all single-particle states for the charge density so we have then:

$$\rho = -e \sum_{n\sigma} |v_{n\sigma}|^2 - e \sum_{\epsilon_{\mathbf{k}}<eV} \sum_\sigma \left(|u_{\mathbf{k}\sigma}|^2 - |v_{\mathbf{k}\sigma}|^2 \right), \tag{7.7}$$

where the first term on the right arises from the ground state. The quasi-particle current density from this generic excited state can also be computed, with the result:

$$
\begin{aligned}
j_y &= -\frac{e}{2m} \sum_{\epsilon_{\mathbf{k}}<eV} \sum_{\sigma} \left\langle -i\psi_\sigma^\dagger \frac{\partial}{\partial y}\psi_\sigma + i\left(\frac{\partial}{\partial y}\psi_\sigma^\dagger\right)\psi_\sigma \right\rangle_{\mathbf{k}} \\
&= -\frac{e}{m}\mathrm{Im}\left[\sum_{n\sigma} v_{n\sigma}\frac{\partial v_{n\sigma}^*}{\partial y} + \sum_{\epsilon_{\mathbf{k}}<eV}\sum_{\sigma}\left(u_{\mathbf{k}\sigma}^*\frac{\partial u_{\mathbf{k}\sigma}}{\partial y} + v_{\mathbf{k}\sigma}^*\frac{\partial v_{\mathbf{k}\sigma}}{\partial y}\right)\right] \quad (7.8) \\
&= -\frac{e}{m}\mathrm{Im}\left[\sum_{\epsilon_{\mathbf{k}}<eV}\sum_{\sigma}\left(u_{\mathbf{k}\sigma}^*\frac{\partial u_{\mathbf{k}\sigma}}{\partial y} + v_{\mathbf{k}\sigma}^*\frac{\partial v_{\mathbf{k}\sigma}}{\partial y}\right)\right],
\end{aligned}
$$

where $\langle...\rangle_{\mathbf{k}}$ is a shorthand notation of $\langle \mathbf{k}|...|\mathbf{k}\rangle$. The first term in the second line of Eq. (7.8) vanishes because it represents the net current for the system in the ground state with a real pair potential. The right hand side of the continuity equation, Eq. (7.5), becomes $-4e\mathrm{Im}\left[\Delta \sum_{\epsilon_{\mathbf{k}}<eV}\left(u_{\mathbf{k}\uparrow}^* v_{\mathbf{k}\downarrow} + v_{\mathbf{k}\uparrow}u_{\mathbf{k}\downarrow}^*\right)\right]$ and is responsible for the interchange between the quasi-particle current density and the supercurrent density [25]. One should always numerically verify, in all computations, that by properly including these terms, all of the numerical results for the current density are constant throughout the whole system. If not, there is a mistake somewhere!

7.3 Looking Ahead

This has been an introduction to transport, in some ways semi-quantitative only. Many things have been promised and most of the details have to be developed. We get on with the details of how to calculate the conductance in Chap. 8. The question of spin transport, which is very important, will be discussed in Chap. 9. Since both charge and spin are carried by the quasiparticles, their transport is not independent, and their interaction will be the subject of Chap. 10.

Chapter 8

Charge Transport

8.1 Introduction

In Chap. 7, we have given a general introduction to the methodology that needs to be used to deal with transport in FS structures. We have stated the general principles, while leaving out most of the technical details. In this Chapter we will fill in these details by discussing the charge transport properties in systems such as the spin valve structure depicted in Fig. 2.1. In doing so, we will make the ideas developed qualitatively in the preceding chapter more rigorous and quantitative.

8.2 Enforcing the Boundary Conditions

As discussed in Sec. 7.1.1 we consider quasiparticles coming from the outer ferromagnet layer F_1, the left side of Fig. 2.1. These quasiparticles can be reflected either by ordinary or by Andreev reflection or can be transmitted as quasiparticles or holes. Since the exchange fields in the F_1 and F_2 layers are, in general, non-collinear, it follows from Eq. (2.8) that the spin-up (-down) quasi-particle wavefunction is not just coupled to the spin-down (-up) quasi-hole wavefunction, as would be the case if only one F layer was present. The wavefunction in the F_1 layer is a linear combination of the original incident spin-up quasi-particle wavefunctions and various types of reflected wavefunctions, namely reflected spin-up and spin-down quasi-particle and quasi-hole wavefunctions. Using the standard column vector notation to represent these combinations, the wavefunctions in F_1 for a spin up particle, will be of the form:

$$\Psi_{F1,\uparrow} \equiv \begin{pmatrix} e^{ik_{\uparrow1}^+ y} + b_{\uparrow,\uparrow} e^{-ik_{\uparrow1}^+ y} \\ b_{\downarrow,\uparrow} e^{-ik_{\downarrow1}^+ y} \\ a_{\uparrow,\uparrow} e^{ik_{\uparrow1}^- y} \\ a_{\downarrow,\uparrow} e^{ik_{\downarrow1}^- y} \end{pmatrix}. \tag{8.1}$$

If the incident particle has spin down, the corresponding wavefunction in F_1 is

$$\Psi_{F1,\downarrow} \equiv \begin{pmatrix} b_{\uparrow,\downarrow} e^{-ik_{\uparrow1}^+ y} \\ e^{ik_{\downarrow1}^+ y} + b_{\downarrow,\downarrow} e^{-ik_{\downarrow1}^+ y} \\ a_{\uparrow,\downarrow} e^{ik_{\uparrow1}^- y} \\ a_{\downarrow,\downarrow} e^{ik_{\downarrow1}^- y} \end{pmatrix}, \tag{8.2}$$

where the $a_{\sigma,\sigma'}$ and $b_{\sigma,\sigma'}$ are reflection and transmission amplitudes, to be determined. One can easily distinguish the physical meaning of each individual wavefunction in this case. For instance in Eq. (8.1), $a_{\downarrow,\uparrow} (0,0,0,1)^T e^{ik_{\downarrow1}^- y}$ is the reflected spin-down quasi-hole wavefunction. The quasi-hole wavefunctions are the time reversed solutions of the BdG equations and carry a positive sign in the exponent for a left-going wavefunction.

As explained in the previous chapter, we must carefully keep track of which are the wavevectors for the particles and holes in the above expressions. I will adopt the notation where $k_{\sigma1}^\pm$ are the quasi-particle $(+)$ and quasi-hole $(-)$ wavevectors in the longitudinal direction y. Fixing the energy to the value ϵ and the transverse wavevector to a value $k_\perp$ (in our quasi one dimensional system this is a conserved quantity) we have the relation:

$$k_{\sigma m}^\pm = \left[\Lambda(1 - \eta_\sigma h_m) \pm \epsilon - k_\perp^2 \right]^{1/2}, \tag{8.3}$$

where $\eta_\sigma = 1$ or -1 for spin down or up, $m = 1$ (as used above) means the first (outer) ferromagnet, and $m = 2$, which we will need later, the inner one. As mentioned in Chap. 2, all energies are in units of E_{FS} and we measure all momenta in units of k_{FS}.

The relevant angles of reflection at the different interfaces can be easily found in terms of wavevector components. Thus, e.g., the incident angle θ_i (for spin-up) at the $F_1 - F_2$ interface (I assume here for the time being that N is not present) is

$$\theta_i = \tan^{-1}\left(k_\perp/k_{\uparrow 1}^+\right), \tag{8.4}$$

while the Andreev reflected angle $\theta_{r\downarrow}^-$ for the reflected spin-down quasi-hole wavefunction is $\theta_{r\downarrow}^- = \tan^{-1}\left(k_\perp/k_{\downarrow 1}^-\right)$. The conservation of transverse momentum leads to many important features when one evaluates the angularly averaged conductance, as we will see below.

To see now how to proceed, consider for example an incoming spin up particle, (Eq. (8.1)) and, forgetting for a while about the existence of the normal and the second ferromagnetic layers, consider the S layer only. If the pair potential was simply stepwise, equal to Δ_0 throughout the S region, we would then have the superconducting coherence factors,

$$u_0 = \frac{1}{\sqrt{2}}\left[\left(\epsilon + \sqrt{\epsilon^2 - \Delta_0^2}\right)/\epsilon\right]^{1/2},$$
$$v_0 = \frac{1}{\sqrt{2}}\left[\left(\epsilon - \sqrt{\epsilon^2 - \Delta_0^2}\right)/\epsilon\right]^{1/2}. \tag{8.5}$$

In this case we would need to consider only the right-going eigenfunctions on the S side, which in general could be written as,

$$\Psi_S \equiv \begin{pmatrix} t_1 u_0 e^{ik^+ y} + t_4 v_0 e^{-ik^- y} \\ t_2 u_0 e^{ik^+ y} + t_3 v_0 e^{-ik^- y} \\ t_2 v_0 e^{ik^+ y} + t_3 u_0 e^{-ik^- y} \\ t_1 v_0 e^{ik^+ y} + t_4 u_0 e^{-ik^- y} \end{pmatrix}, \tag{8.6}$$

where now

$$k^\pm = \left[1 \pm \sqrt{\epsilon^2 - \Delta_0^2} - k_\perp^2\right]^{1/2} \tag{8.7}$$

would be quasi-particle (+) and quasi-hole (−) wavevectors in the S region. I should have added a spin index to the t_i transmission amplitudes, since they do depend on whether the incoming spin is up or down, but the notation is messy enough already.

8.3 Extracting the Conductance

We can now see what is to be done to obtain the conductance: in Eq. (8.6) there are four unknown coefficients. There are four more unknown coefficients in Eq. (8.1). We have eight equations: continuity of the four component wave function at the interface and discontinuity of the derivative, as determined by interfacial δ function scattering. That's it. We can get the normal and Andreev reflection coefficients and substitute the result in Eq. (7.2). The wavevector ratios in that equation are now known: we have written the wavevectors down in Eq. (8.7) and in general in Eq. (8.3).

How about the F_2 and the N layers, which we have not yet included? This is not difficult. In these layers we will have both right moving and left moving particles and holes. So, instead of four new coefficients as in Eq. (8.6) we would have eight. But for each new layer we will have an additional interface. Associated with it there will be eight more continuity conditions. Therefore while the system of equations will get bigger and more complicated, the number of equations and unknowns will grow by the same number.

Of course, we have not yet dealt with the spatial dependence of the self-consistent pair potential, and we know that it is *very* important because of the conservation laws, as seen in Sec. 7.2. We will do this in the next subsection, but first we want to write the general form of the wavefunction in F_2 remembering that, in general, the internal fields of the two ferromagnets are rotated by an angle ϕ. Specifically, then, in the intermediate layer F_2, we can write:

$$\Psi_{F2} \equiv \begin{pmatrix} c_1 f_\uparrow^+ e^{ik_{\uparrow 2}^+ y} + c_2 f_\uparrow^+ e^{-ik_{\uparrow 2}^+ y} + c_3 g_\uparrow^+ e^{ik_{\downarrow 2}^+ y} + c_4 g_\uparrow^+ e^{-ik_{\downarrow 2}^+ y} \\ c_1 f_\downarrow^+ e^{ik_{\uparrow 2}^+ y} + c_2 f_\downarrow^+ e^{-ik_{\uparrow 2}^+ y} + c_3 g_\downarrow^+ e^{ik_{\downarrow 2}^+ y} + c_4 g_\downarrow^+ e^{-ik_{\downarrow 2}^+ y} \\ c_5 f_\uparrow^- e^{ik_{\uparrow 2}^- y} + c_6 f_\uparrow^- e^{-ik_{\uparrow 2}^- y} + c_7 g_\uparrow^- e^{ik_{\downarrow 2}^- y} + c_8 g_\uparrow^- e^{-ik_{\downarrow 2}^- y} \\ c_5 f_\downarrow^- e^{ik_{\uparrow 2}^- y} + c_6 f_\downarrow^- e^{-ik_{\uparrow 2}^- y} + c_7 g_\downarrow^- e^{ik_{\downarrow 2}^- y} + c_8 g_\downarrow^- e^{-ik_{\downarrow 2}^- y} \end{pmatrix}, \quad (8.8)$$

where $k_{\uparrow 2}^\pm$ and $k_{\downarrow 2}^\pm$ are defined in Eq. (8.3) (that was the reason for the $m = 2$ subscript there). The $\pm$ indices are defined as previously, and the up and down arrows refer to the spin axes in F_1. The eigenspinors f and g are needed because of the spin rotation required to go from the spin direction associated with F_1 to F_2. They correspond to spin parallel or antiparallel to $\mathbf{h_2}$ respectively. They are given, for $0 \leq \phi \leq \pi/2$, by the expression,

$$\begin{pmatrix} f_\uparrow^+ \\ f_\downarrow^+ \end{pmatrix} = \frac{1}{\mathcal{N}} \begin{pmatrix} 1 \\ \dfrac{1 - \cos\phi}{\sin\phi} \end{pmatrix} = \begin{pmatrix} f_\uparrow^- \\ -f_\downarrow^- \end{pmatrix};$$

$$\begin{pmatrix} g_\uparrow^+ \\ g_\downarrow^+ \end{pmatrix} = \frac{1}{\mathcal{N}} \begin{pmatrix} -\dfrac{\sin\phi}{1 + \cos\phi} \\ 1 \end{pmatrix} = \begin{pmatrix} -g_\uparrow^- \\ g_\downarrow^- \end{pmatrix}$$

$$(8.9)$$

with the normalization constant $\mathcal{N} = \sqrt{2/1 + \cos\phi}$. These spinors reduce to those for pure spin-up and spin-down quasi-particles and holes when $\phi = 0$, corresponding to a uniform magnetization along z. One can also easily see that the particular wavefunction of Eq. (8.8), $c_1 \left(f_\uparrow^+, f_\downarrow^+, 0, 0 \right)^T e^{ik_{\uparrow 2}^+ y}$, denotes a quasi-particle with spin parallel to the exchange field in F_2. When $\pi/2 < \phi \leq \pi$, these eigenspinors read

$$\begin{pmatrix} f_\uparrow^+ \\ f_\downarrow^+ \end{pmatrix} = \frac{1}{\mathcal{N}} \begin{pmatrix} \sin\phi \\ 1 - \cos\phi \\ 1 \end{pmatrix} = \begin{pmatrix} -f_\uparrow^- \\ f_\downarrow^- \end{pmatrix};$$

$$\begin{pmatrix} g_\uparrow^+ \\ g_\downarrow^+ \end{pmatrix} = \frac{1}{\mathcal{N}} \begin{pmatrix} 1 \\ 1 + \cos\phi \\ -\sin\phi \end{pmatrix} = \begin{pmatrix} g_\uparrow^- \\ -g_\downarrow^- \end{pmatrix}$$

$$(8.10)$$

with $\mathcal{N} = \sqrt{2/1 - \cos\phi}$.

8.3.1 *Self-consistency: General*

The above is basically all we need, except of course for the pesky question of enforcing self-consistency, which requires dealing with the spatial dependence of the pair potential. The obtention of the self-consistent pair potential $\Delta(y)$ proceeds exactly as in the basic formalism in Chap. 2. It is true that the geometry is more complicated, but formally this simply amounts to a more complicated y dependence of the fields $\mathbf{h}(y)$ and messier matrix elements. There is therefore no need to repeat the formalism except to point out that since one now must deal with complex functions; it is convenient to use plane waves instead of sinc functions as the expansion functions in the procedure described in Sec. 2.5.1, (see Eq. (2.14)). This poses no complications and changes nothing in the equations except for the detail of adding the appropriate complex conjugates in the right places. For example, the self-consistency condition, Eq. (2.17), now becomes:

$$\Delta(y) = \frac{g(y)}{2} \sum_n{}' \left[u_{n\uparrow}(y) v_{n\downarrow}^*(y) + u_{n\downarrow}(y) v_{n\uparrow}^*(y) \right] \tanh\left(\frac{\epsilon_n}{2T} \right). \qquad (8.11)$$

Of course, with plane waves one employs appropriate periodic boundary conditions. As long as there is only one superconducting layer, there is an overall arbitrary phase factor. This is not the case for Josephson structures, as studied in Chaps. 11 and 12. Furthermore, if one uses the transfer matrix procedure discussed below in Sec. 8.3.2 to extract the conductance it is not strictly necessary to use plane waves in the preliminary step. However, to study spin transport the complex functions are needed.

8.3.2   *Self-consistency: The transfer matrix method*

When one gets the self-consistent pair potential, by repeated diagonalization, one obtains also the wave functions. One can then in principle look for linear combinations of these wave functions which satisfy the boundary conditions and then extract the conductance. This is somewhat awkward, but quite feasible. There is a much easier way, however, and it is to start with the known complete set of wave functions and then use a transfer matrix [28] procedure to find all needed coefficients, including of course the reflection amplitudes need to evaluate the conductance G.

A non-self-consistent step potential assumption is of course very unrealistic, even leaving aside the lack of self-consistency. Proximity effects lead to a complicated behavior of the superconducting order parameter in the F layers, studied in the simplest case in Sec. 2.7 (see e.g., Fig. 2.4). It leads also to the generation of triplet pairs as discussed in Chap. 3. The concomitant depletion of the pair amplitudes near the $F - S$ interface means that unless the superconductor is thick enough, the pair amplitude does not saturate to its bulk value even deep inside the S regions. Furthermore, as we have seen in Sec. 7.2, lack of self-consistency may lead to violation of charge conservation: hence, while non-self-consistent approximations might sometimes be adequate for equilibrium calculations, their use for transport should be viewed as acceptable for illustrative and pedagogical purposes only, and otherwise be eschewed. Therefore, one should generally use a self-consistent pair potential that is allowed to vary spatially, as obtained by using Eq. (2.12), and which results in a minimum in the free energy of the system.

Let us begin by extending the BTK formalism to the spatially varying self-consistent pair potential obtained as explained above, enforcing Eq. (2.12). In this way we can include the influence of the superconducting proximity effects on the tunneling conductance. Although the self-consistent solutions of the BdG equations reveal that the pair *amplitudes* are non-zero in the non-superconducting regions due to the proximity effects, the pair *potential* itself vanishes in these regions since $g(y) \equiv 0$ there. Therefore, one can still use Eqs. (8.1) and (8.2), together with Eq. (8.8), for the wavefunctions in the F_1 and F_2 regions. As far as the N layer goes, it is the same as an additional intermediate S layer, with zero internal field of course. In the discussion of this transfer matrix method I will largely omit mentioning the normal layer, as its inclusion is trivial.

So, the only difficult thing is to deal with the spatially varying pair potential on the S layer. To do so we divide it into many very thin layers with microscopic thicknesses of order k_{FS}^{-1}. We can treat each layer as a very thin superconductor with a constant pair potential, Δ_i, as obtained from the self-consistent procedure. We are then able to write the eigenfunctions of each superconducting layer corresponding to that value of the pair potential. For example, in the i-th layer, the eigenfunction should contain all left and right going solutions, and it reads:

$$
\Psi_{S_i} \equiv
\begin{pmatrix}
t_{1i} u_i e^{ik_i^+ y} + \tilde{t}_{1i} u_i e^{-ik_i^+ y} + t_{4i} v_i e^{-ik_i^- y} + \tilde{t}_{4i} v_i e^{ik_i^- y} \\
t_{2i} u_i e^{ik_i^+ y} + \tilde{t}_{2i} u_i e^{-ik_i^+ y} + t_{3i} v_i e^{-ik_i^- y} + \tilde{t}_{3i} v_i e^{ik_i^- y} \\
t_{2i} v_i e^{ik_i^+ y} + \tilde{t}_{2i} v_i e^{-ik_i^+ y} + t_{3i} u_i e^{-ik_i^- y} + \tilde{t}_{3i} u_i e^{ik_i^- y} \\
t_{1i} v_i e^{ik_i^+ y} + \tilde{t}_{1i} v_i e^{-ik_i^+ y} + t_{4i} u_i e^{-ik_i^- y} + \tilde{t}_{4i} v_i e^{ik_i^- y}
\end{pmatrix},
\qquad (8.12)
$$

where, again, I have omitted spin indices,

$$
k_i^\pm = \left[1 \pm \sqrt{\epsilon^2 - \Delta_i^2} - k_\perp^2 \right]^{1/2}
\qquad (8.13)
$$

and Δ_i represents the strength of the normalized self-consistent pair potential in the i-th superconducting layer. The superconducting coherence factors u_i and v_i depend on Δ_i in the standard way, as in Eq. (8.5) but with Δ_i replacing Δ_0. Of course there are no scattering barriers between these layers, called into being merely for mathematical convenience. All the coefficients in Eq. (8.12) are unknown, and remain to be determined. However, in the outermost S layer (rightmost in our convention) the eigenfunctions contains only right-going terms, that is, it is of a form identical to Eq. (8.6) but with different locally constant pair potential.

We see then that the price one has to pay for including the proximity effects is the need to compute a very large number of coefficients. The same counting arguments used in the previous subsection reveal that the number of unknowns equals that of equations, but the number of equations is of course extremely large. Here is where transfer matrix methods to solve for these unknowns [28] come handy. If one considers the interface between the i-th and the $(i+1)$-th layer, the continuity conditions will give a linear relation between the coefficients:

$$
\tilde{\mathcal{M}}_i x_i = \mathcal{M}_{i+1} x_{i+1},
\qquad (8.14)
$$

where, for a generic i one has,

$$x_i^T \equiv \left(t_{1i}, \tilde{t}_{1i}, t_{2i}, \tilde{t}_{2i}, t_{3i}, \tilde{t}_{3i}, t_{4i}, \tilde{t}_{4i} \right), \tag{8.15}$$

and the matrices, $\tilde{\mathcal{M}}_i$ and $\mathcal{M}_{i+1}$, can be written from the continuity conditions as discussed above. Conversely the coefficients in the $(i+1)$-th layer can be obtained in terms of those in the i-th layer as

$$x_{i+1} = \mathcal{M}_{i+1}^{-1} \tilde{\mathcal{M}}_i x_i. \tag{8.16}$$

In the same way, for the interface between the $(i-1)$-th layer and the i-th layer, we can write

$$x_i = \mathcal{M}_i^{-1} \tilde{\mathcal{M}}_{i-1} x_{i-1}. \tag{8.17}$$

From the above relations, one can write down the relation between x_{i+1} and x_{i-1}, i.e.,

$$x_{i+1} = \mathcal{M}_{i+1}^{-1} \tilde{\mathcal{M}}_i \mathcal{M}_i^{-1} \tilde{\mathcal{M}}_{i-1} x_{i-1}. \tag{8.18}$$

By iteration of this procedure, one can "transfer" the coefficients layer by layer and eventually relate the coefficients of the rightmost layer, x_n, where, we recall, there are only four coefficients per spin instead of the usual eight, to those of the leftmost (innermost) layer in S. One can then transfer on to the inner ferromagnetic layer F_2, either via an intermediate step at the $N - F_2$ interface, or, if N is absent, directly:

$$x_n = \mathcal{M}_n^{-1} \tilde{\mathcal{M}}_{n-1} \mathcal{M}_{n-1}^{-1} \cdots \tilde{\mathcal{M}}_1 \mathcal{M}_1^{-1} \tilde{\mathcal{M}}_{F2} x_{F2}. \tag{8.19}$$

By solving Eq. (8.19) together with $\mathcal{M}_{F1} x_{F1} = \mathcal{M}_{F2} x_{F2}$, we obtain all the coefficients in the F_1 region, where the wavefunction is *formally* still described by the expressions given in Eqs. (8.1) and (8.2). Of course, all coefficients involved, including the energy dependent $a_{\sigma\sigma'}$ and $b_{\sigma\sigma'}$ values from which the conductance is extracted, are quite different from those in a non-self-consistent calculation. These differences are reflected in the results. One can also prove that, when the pair potential in S is taken to be a constant (non-self-consistent), then $\mathcal{M}_{i+1} = \tilde{\mathcal{M}}_i$ and therefore Eq. (8.19) becomes $x_n = x_1 = \mathcal{M}_1^{-1} \tilde{\mathcal{M}}_{F2} x_{F2}$. This can be seen to be formally identical to that we have seen in our discussion of the non-self-consistent formalism in Sec. 8.2.

This efficient technique, besides allowing the determination of all the reflected and transmitted amplitudes in the outermost layers, permits performing a consistency check by recomputing the self-consistent eigenfunctions of the BdG equations. Once one has determined the amplitudes x_{F1}, x_{F2}, and x_n, one can use them to find the amplitudes in any intermediate layer by "transferring" back the solutions. For example, the coefficients x_{n-1} can be found by using $x_n = \mathcal{M}_n^{-1}\tilde{\mathcal{M}}_{n-1}x_{n-1}$ if one knows the coefficient x_n for the rightmost layer. Knowledge of these coefficients in every region yields again the self-consistent wavefunctions of the system. These of course should be the same as the eigenfunctions found in the original procedure. Although the numerical computations involved in this consistency check are rather intensive, it is possible to perform them: one can verify that, by plugging these solutions into Eq. (2.12) and considering all possible solutions with all possible incident angles to the BdG equations, the output pair potential obtained from the transport calculation is the same as the input pair potential obtained by direct diagonalization. This would obviously not have been the case if the initial pair potential had not been fully self-consistent to begin with.

The differences between the conductance obtained from the self-consistent calculation of the reflected and transmitted amplitudes and that obtained from the non-self-consistent calculation are quite important, as we shall discuss later.

Having done all this, we are in a position to write our final expression for the conductance contribution for particles with energy ϵ and entering the sample at an angle θ_i. Namely, from the above arguments it follows that:

$$
G(\epsilon, \theta_i) = \sum_\sigma P_\sigma G_\sigma(\epsilon, \theta_i)
$$

$$
= \sum_\sigma P_\sigma \left(1 + \frac{k_{\uparrow 1}^-}{k_{\sigma 1}^+}|a_{\uparrow\sigma}|^2 + \frac{k_{\downarrow 1}^-}{k_{\sigma 1}^+}|a_{\downarrow\sigma}|^2 - \frac{k_{\uparrow 1}^+}{k_{\sigma 1}^+}|b_{\uparrow\sigma}|^2 - \frac{k_{\downarrow 1}^+}{k_{\sigma 1}^+}|b_{\downarrow\sigma}|^2 \right),
$$

$$(8.20)$$

where, as already mentioned, the conductance is in its natural (e^2/h) (see Table 2.1) units. In the above expression the meaning of the different k ratios has been explained in connection with Eq. (7.2) and the relevant wavevectors are defined above (see e.g., Eqs. (8.3) and (8.7)). The $a_{\sigma\sigma'}$ and $b_{\sigma\sigma'}$ amplitudes are as defined in Eqs. (8.1) and (8.2). These coefficients, which are, to repeat, energy-dependent, are calculated using the self-consistent transfer matrix technique of Sec. 8.3.2. Therefore, even though

Eq. (8.20) is formally the same in the self-consistent and non-self-consistent cases, the results for the reflection amplitudes or probabilities involved, $|a_{\sigma\sigma'}|^2$, and $|b_{\sigma\sigma'}|^2$, are different in these two schemes. The angle θ_i is the incident angle, given in terms of $\mathbf{k}$ components by Eq. (8.4). The weight factors, $P_\sigma \equiv (1 - h_1\eta_\sigma)/2$ account for the number of available states for spin-up and spin-down bands in the outer electrode.

It is also possible, but computationally more intricate, to evaluate the current density directly from the wavefunctions, and one can somewhat painfully verify that the resulting current density is identical to the terms inside the bracket in the expression of G, Eq. (8.20). In other words, in the low-T limit the two methods are equivalent. We will discuss the effects of temperature later. Also, it is well to anticipate here, that no convenient transfer matrix shortcut seems to exist to calculate spin currents, and that one must use then the expressions that one can find in terms of the wavefunctions.

8.3.3 *Angles and angular averages*

The conductance results, Eq. (8.20), obviously depend on the incident angle of electrons, θ_i. Experimentally, one frequently measures simply the forward conductance, $\theta_i = 0$, via point contacts or, in other experimental conditions, an appropriate angular average. Consequently, it is worthwhile to discuss here the angularly averaged conductance by using the following definitions,

$$\langle G_\sigma(\epsilon)\rangle = \frac{\int_0^{\theta_{c\sigma}} d\theta_i \cos\theta_i G_\sigma(\epsilon,\theta_i)}{\int_0^{\theta_{c\sigma}} d\theta_i \cos\theta_i}, \tag{8.21}$$

and

$$\langle G\rangle = \sum_\sigma P_\sigma \langle G_\sigma\rangle, \tag{8.22}$$

where the critical angle $\theta_{c\sigma}$ is in general different for spin-up and spin-down bands. This is because the critical angle arises from the conservation of transverse momentum and the corresponding Snell's law:

$$\sqrt{\left(k_{\sigma 1}^+{}^2 + k_\perp^2\right)}\sin\theta_i = \sqrt{\left(k_{\sigma' 1}^+{}^2 + k_\perp^2\right)}\sin\theta_{r\sigma'}^+$$

$$= \sqrt{\left(k_{\sigma' 1}^-{}^2 + k_\perp^2\right)}\sin\theta_{r\sigma'}^- = \sin\theta_S, \tag{8.23}$$

where we recall again that we measure wavevectors in units of k_{FS}. The angles $\theta^{\pm}_{r\sigma}$ equal $\tan^{-1}\left(k_{\perp}/k^{\pm}_{\sigma 1}\right)$, and the σ and σ' are each $\uparrow$ or $\downarrow$. The last equality in Eq. (8.23) represents the case of the transmitted wave in S, and θ_S is the transmitted angle. Although the self-consistent pair potential varies within S and so do the quasi-particle (hole) wavevectors, we here need only consider the transmitted angle θ_S in the rightmost layer: this follows in the same way as the usual Snell's law in a layered system, as given in elementary textbooks. From Eq. (8.23), one can determine the critical angles for different channels. Consider, e.g., a spin-up electron incident from F_1 without any Fermi wavevector mismatch, i.e., $\Lambda = 1$. Since we are only concerned (see Sec. 7.1.2) with the case that the bias of tunneling junctions is of the order of superconducting gap and therefore much smaller than the Fermi energy, the approximate magnitude of the incident wavevector is $\sqrt{1+h_1}$: this is called the Andreev approximation. One can substitute this and similar expressions into Eq. (8.23) and, with the help of Eq. (8.3), obtain

$$\sqrt{1+h_1}\sin\theta_i = \sqrt{1-h_1}\sin\theta^-_{r\downarrow} = \sin\theta_S. \qquad (8.24)$$

One can straightforwardly verify that, when the relation $\theta_i > \sin^{-1}[((1-h_1)/(1+h_1))^{1/2}]$ is satisfied for the incident angle, the conventional Andreev reflection becomes an evanescent wave. In this case, the conventional Andreev reflection does not contribute to the angular averaging. On the other hand, if the energy ϵ of the incident electron is less than the saturated value of the superconducting pair amplitude in S, all the contribution to the conductance from the transmitted waves in S also vanishes because $k^{\pm}$ acquires an imaginary part. However, even the condition that ϵ is greater than the saturated superconducting amplitude does not guarantee that the contribution from the transmitted waves to the conductance is nonvanishing. One still needs to consider the transmitted critical angle $\sin^{-1}[1/(1+h_1)^{1/2}]$. So one most define an overall critical angle $\theta_{c\sigma}$ as the largest one among all the reflected and transmitted critical angles. It is obvious that the critical angles $\theta_{c\sigma}$ are different for spin-up and spin-down bands when $h_1 \neq 0$.

In many cases angularly averaged results are not needed to interpret experimental results. This is the case for point contact experiments, that involve only the forward conductance. Hence in the rest of this book the forward conductance G will be emphasized, and the word "conductance" will by default refer to G. Whenever angular averages are presented, it will be made clear and the notation $\langle G \rangle$ will be employed.

8.4 Conductance Results

8.4.1 *General considerations*

We now start a discussion of conductance results obtained using the methods that we have just discussed. First, we will discuss FFS trilayer structures, with some preliminary examples involving FS bilayers. Next, beginning with Sec. 8.4.4, we will consider full realistic spin valve structures including the intermediate normal layer, as in Fig. 2.1. The figures in this section are taken from Refs. [29] (Figs. 8.1–8.9 and [30] (Figs. 8.10–8.15). The results presented are mostly for the forward conductance but angularly averaged results are also given, particularly in the context of discussing the different kinds of Andreev reflection (see Sec. 8.4.3). The bias dependent conductances $G(E)$ are computed by considering a particle incident with an angle $\theta_i \cong 0$ (normal incidence). Recall that we normalize the bias to the zero temperature bulk S gap Δ_0, $E \equiv eV/\Delta_0$, as explained in Sec. 7.1.2. Angular averaging has been discussed in the text above Eq. (8.21). G itself is normalized to the conductance quantum (see Table 2.1).

For structures where the F_1 and F_2 regions are made of the same ferromagnetic material, i.e., $h_1 = h_2$ and $k_{\uparrow 1,(\downarrow 1)}^{\pm} = k_{\uparrow 2,(\downarrow)2}^{\pm}$, I will use h (not to be confused with Planck's constant) and $k_{\uparrow,(\downarrow)}^{\pm}$ to denote their exchange fields and wavevectors. This is the case we will mostly study since it is the one that is experimentally relevant, see e.g., Chap. 6: it is hard enough to make good samples using one magnetic material, so it seems futile to use two. Only very occasionally (in connection with Fig. 8.5) will we consider dissimilar magnetic materials. Results are for the low-T limit except for the explicit discussion of the temperature dependence. As in other chapters, the lengths are measured in units of k_{FS}^{-1} and denoted by capital letters, e.g., D_S denotes $k_{FS}d_S$.

Many acronyms are used in the discussion, both in this chapter and elsewhere later. They are all listed in Table 8.1.

8.4.2 *FFS trilayers: Forward conductance*

In this subsection we want to see the conductance results for FFS trilayers, leaving for later the inclusion of the necessary N layer spacer. Let us begin with a brief introductory discussion (Fig. 8.1) contrasting self-consistent and non-self-consistent (step function pair potential) results for the forward tunneling conductance in FS bilayers. In this figure the dimensionless superconducting coherence length Ξ_0 is taken to be 50 and the thicknesses

Table 8.1 Acronyms used in the transport chapters.

Acronym	Meaning
CB	Critical Bias
ZBC	Zero Bias Conductance
CBC	Critical Bias Conductance
ZBCP	Zero Bias Conductance Peak
ZBCD	Zero Bias Conductance Dip
FBCP	Finite Bias Conductance Peak
P, AP	Parallel, Antiparallel
ESAR (AAR)	Equal Spin (Anomalous) Andreev Reflection
STT	Spin Transfer Torque
AR	Andreev Reflection
MAR	Multiple Andreev Reflection

D_F and D_S of the F and S layers are both $15\Xi_0$. Therefore the S layer is thick enough (a large ratio of D_S to Ξ_0) so that one expects that the pair amplitude saturates to its bulk value deep inside it. This is indeed the case, as can be verified by computing the pair amplitudes via the direct diagonalization method, Chap. 2. Because of this, for these large sample thicknesses, the proximity effects are less important. The situation is, then, the least unfavorable for the non-self-consistent approximation.

The replacement in NS bilayers, of the non-magnetic metal with a ferromagnet leads to suppression of the Andreev reflection in the region that we call (Sec. 7.1.2) the subgap region. This suppression is strong when the internal field is large: this is because an incoming electron in the majority spin band (which will be the larger contributor to G) will be Andreev reflected as hole in the other band, and there are just fewer available states there. The decrease of the zero bias conductance (ZBC) strongly depends on the magnitude of the exchange field in F. This dependence can be and is used to measure the degree of spin-polarization of magnetic materials experimentally. However, to accurately determine the degree of spin-polarization, one has to consider the Fermi wavevector mismatch (FWM), Λ, as well as the interfacial barriers. The ZBC peak is very sensitive to both spin-polarization and Λ and the dependence cannot be characterized by a single parameter.

In this Fig. 8.1, the results are plotted at two different values of the exchange fields and several wavevector mismatch Λ values. One sees at once that the self-consistent results (dotted lines) approach in this case the non-self-consistent ones (solid lines) in the zero bias limit, while deviating the most for energies near the superconducting gap. The ZBC decreases

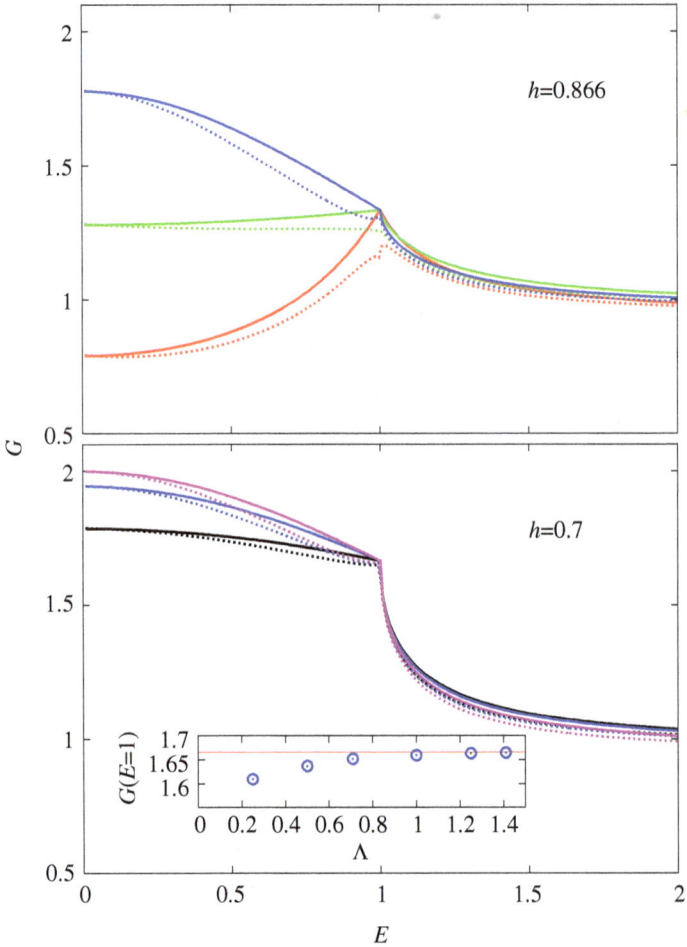

Fig. 8.1 The forward conductance, $G(E)$, in thick FS bilayers (see text). Values of h are indicated. In both main panels solid and dashed curves show G, for non-self-consistent and self-consistent results, respectively. In the top panel the lower (red) curves at low bias are for $\Lambda = 0.25$, the middle (green) curves for $\Lambda = 0.5$, and the higher (blue) curves for $\Lambda = 1$. In the bottom panel, the top (purple) curves are for $\Lambda = 1.41$, the middle (blue) curves are as in the top panel, and the lower (black) ones for $\Lambda = 0.71$. The inset shows $G(E = 1)$ vs. Λ in the self-consistent calculation (dots) and the non-self-consistent result (line).

with increasing h and with decreasing Λ. Also, larger h indeed leads to a conspicuous reduction in the subgap conductance, for the reasons already mentioned, and so does the introduction of FWM. One can conclude that, as mentioned above, the behavior of the ZBC cannot be characterized by only one parameter, either h or Λ alone. Instead, one should expand the fitting parameter space to determine the degree of spin polarization.

In the non-self-consistent approximation, the conductance at the superconducting gap ($E = 1$ in our units) is independent of Λ at a given h. However, this conclusion is invalid in the correct self-consistent approach, and the conductance at the superconducting gap varies in fact monotonically with increasing Λ. The inset in the bottom panel of Fig. 8.1 clearly shows this dependence on Λ. Figure 8.1 also shows that the self-consistent results (dashed curves) on subgap conductances are in general lower than those obtained in the non-self-consistent framework (solid curves) for a strong exchange field. On the other hand, in the high bias limit, the self-consistent results become similar to the non-self-consistent ones. This is simply because the particle does not experience much of a difference between a step-like pair potential and a smooth pair potential when it is incident with high enough energy, $E \gg 1$. Finally, clear cusps appear at the superconducting gap value in some cases, e.g., the forward scattering conductance curve at $h = 0.866$ and $\Lambda = 1$.

We can now discuss in a similar way results for $F_1 F_2 S$ trilayers of experimentally relevant widths. First, we consider the dependence of G on the angle ϕ between $\mathbf{h}_1$ and $\mathbf{h}_2$ (see Fig. 2.1). An important feature of these trilayers is the strong dependence of the superconducting transition temperatures T_c on the angle ϕ. This is due to proximity effects and induced long-range triplet correlations. Field-induced switching effects also make these structures attractive candidates for memory elements. The non-monotonic behavior of $T_c(\phi)$, with its minimum being near $\phi = 90°$, has been extensively discussed in Chap. 6. We saw there that this angular dependence is related to the induced (see Sec. 6.3.2) triplet pairing correlations. The superconducting transition temperatures should also be positively correlated with the singlet pair amplitudes deep inside the S regions. Therefore, it is of particular importance to consider systems of finite size to take into view the whole picture of proximity effects on the angular dependence of the tunneling conductance.

For the results shown here in this context, and to isolate the influence of the effects discussed at the end of the previous paragraphs, I will assume the absence of FWM ($\Lambda = 1$). Interfacial scattering is for the time being

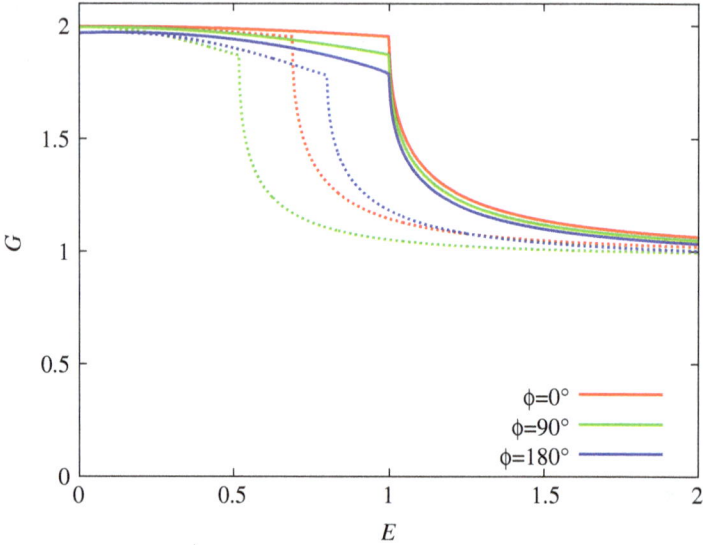

Fig. 8.2 Comparison between self-consistent and non-self-consistent forward scattering conductances of F_1F_2S trilayers. The solid and dashed lines are for non-self-consistent and self-consistent results respectively. The solid curves highest (red) at the critical bias (CB) are for $\phi = 0°$, and those lowest (blue) at CB, for $\phi = 180°$. We have $D_{F1} = 10$, $D_{F2} = 12$, and $D_S = 180$. The internal field is $h = 0.3$ in both layers.

suppressed; its effects are discussed below in Sec. 8.4.5. As a typical example, we see in Fig. 8.2 results for the ϕ dependence of the forward scattering conductances. The exchange field used there for both F layers is $h = 0.3$, and the thicknesses of the F_1 and F_2 layers correspond to $D_{F1} = 10$ and $D_{F2} = 12$ respectively, while the S layer has width $D_S = 180 = 1.5\Xi_0$. Results obtained via the non-self-consistent approach are plotted for comparison. In this more realistic case the differences between the two methods are of course much larger. In the non-self-consistent framework where the single parameter Δ_0 describes the stepwise pair potential, one sees in Fig. 8.2 that for all values of the angle ϕ the conductance curves drop when the bias is at Δ_0, corresponding to $E = 1$ in our units. In contrast, for the self-consistent results, one can clearly see in Fig. 8.2, that the drop in the conductance curves occurs at different bias values for different angles. We also see that this critical bias (which we denote by CB, as in Table 8.1) depends on ϕ non-monotonically, with $\phi = 180°$ corresponding to the largest and $\phi = 90°$ to the smallest CB values. Since the CB depends on the strength of the superconducting gap deep inside the S

regions, the non-monotonicity of the CB in Fig. 8.2 is correlated with the non-monotonicity of T_c. The CB never reaches unity, in these trilayers: because of their finite size the pair potential is mostly below its saturated value. Accordingly, this feature of the correct self-consistent results implies that one cannot adequately determine the angular dependence of the forward conductance in the non-self-consistent framework. This feature also provides experimentalists with another way to measure the strength of the superconducting gap for different angles in these trilayers by determining the CB in a set of conductance curves.

The remaining results shown in this chapter are all computed self-consistently, the above two examples being sufficient for the purpose of showing some of the differences.

In Fig. 8.3, there are more results for the dependence of the forward scattering conductances on ϕ. In the top panels the thicknesses of each layer and the coherence length are the same as for Fig. 8.2. In the bottom panels the thickness of the inner magnetic layer is increased to $D_{F2} = 18$ while D_{F1}, D_S, and Ξ_0 remain unchanged. For each row of Fig. 8.3, results for three different exchange fields are plotted. In the top left panel ($h = 0.5$) we see that the angular dependence of the CB (or the related magnitude of the saturated pair amplitudes) is monotonic with ϕ. Although this monotonicity is not common, one can verify that it is consistent with the theoretical results for $T_c(\phi)$ for the same particular case. The more usual non-monotonic dependence is found in all other panels, as discussed in the previous paragraph. In every case, it turns out that the magnitude of the CB reflects the magnitude of the self-consistent pair amplitudes deep inside the superconductor.

For the ZBC, the degree of its angular dependence is clearly very sensitive to h. In the top left panel, with $h = 0.5$, the ZBC is nearly independent of ϕ. On the other hand, the ZBC in the top right panel, $h = 0.6$, drops by almost a factor of two as ϕ varies from the relative parallel (P) orientation, $\phi = 0°$, to the antiparallel (AP) orientation, $\phi = 180°$. This is a consequence of interference between the spin-up and spin-down wavefunctions under the influence of the rotated exchange field in the middle layer. We will see some more details below and also in Chap. 10. In the top left panel, one sees that the conductance at CB decreases with increasing angle. In other words, the zero bias conductance peak (ZBCP) becomes more prominent as ϕ is increased. However, for the top middle panel, $h = 0.45$, the development of the ZBCP is less noticeable when the angle ϕ is increased. In the top right panel, $h = 0.6$, the ZBCP evolves into a zero bias conductance dip

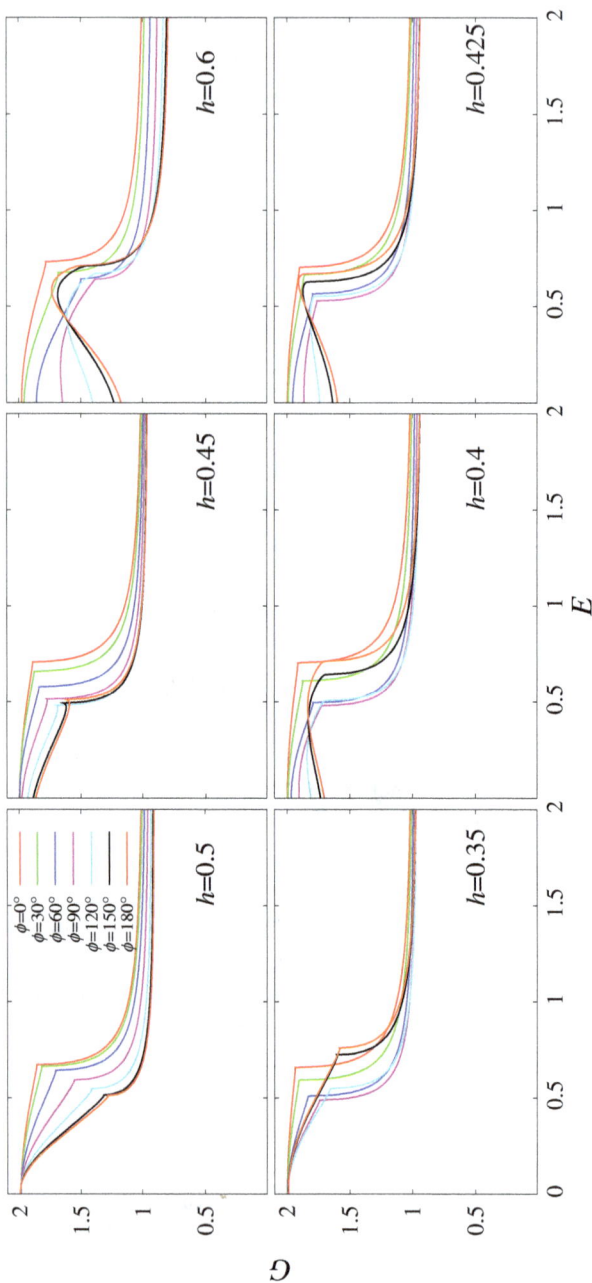

Fig. 8.3 Forward scattering conductance of $F_1 F_2 S$ trilayers for several angles ϕ as indicated in the legend. The top panels are for $D_{F1} = 10$, $D_{F2} = 12$, and $D_S = 180$ and the bottom panels for $D_{F1} = 10$, $D_{F2} = 18$, and $D_S = 180$. The exchange field strength h is indicated. For the two panels on the left, the conductances at CB decrease with increasing ϕ. For the other four panels, the ZBC (see text) decreases as ϕ increases.

(ZBCD) as ϕ varies from $\phi = 0°$ to $\phi = 180°$, with a clear finite bias conductance peak (FBCP) appearing just below the CB. In the bottom panels of this figure, corresponding to a larger value of D_{F2} one can observe similar features. For example, a slight change from $h = 0.35$ to $h = 0.4$ causes by itself a very large change in the behavior of the ZBC. Moreover, the evolution of the ZBCP to a ZBCD accompanies the occurrence of a FBCP when $\phi > 90°$. The location of the FBCP also moves closer to the CB value when ϕ increases. That these features of the ZBC depend on both the strength of exchange field (reflected in $k_\uparrow^\pm$ and $k_\downarrow^\pm$ as defined in Eq. (8.3)) and the thickness of the F_2 layer indicates that the ZBC shows the characteristics of a resonance scattering phenomenon as in an elementary quantum mechanical barrier. The main difference is that the scattering problem here involves the intricate interference between quasi-particle and quasi-hole spinors.

8.4.2.1 *Magnetoresistance of FFS structures*

When the bias is high enough, the tunneling conductance approaches its normal state value. Thus, one can extract the magnetoresistance from the conductance at, say, $E = 2$. I do not want to have any extensive discussion of the magnetoresistance here, since at these large bias values it is rather peripheral to the emphasis of this book, namely the proximity effects. So I will discuss here only briefly the magnetoresistance's qualitative behavior. One can define a measure of the magnetoresistance as,

$$M_G(E, \phi) \equiv \frac{G(E, \phi = 0°) - G(E, \phi)}{G(E, \phi = 0°)}, \tag{8.25}$$

where ϕ is the misalignment angle between the two ferromagnets. In G, it should not be confused with the incident angle used for example in Eq. (8.20). For all results shown in the panels of Fig. 8.3, the conductance at $E = 2$ decreases with increasing ϕ, i.e., $M_G(E = 2, \phi)$ is a monotonic function of ϕ, the standard behavior for conventional, non-superconducting, spin-valves. It is worth reminding the reader here that, generally, a spin valve is a structure that has a different conductance for up and down spins. Furthermore, one can also see that $M_G(E = 2, \phi = 180°)$ increases with exchange field. Therefore, the behavior of the magnetoresistance at large bias is as one would expect in the present self-consistent framework. However, the behavior of $M_G(E = 0, \phi = 180°)$ that is associated with the behavior of the ZBC is generally a non-monotonic function of h.

8.4.2.2 *Resonances in the zero bias conductance*

It is interesting to consider the reasons for the high sensitivity of the ZBC to h by examining its resonances for two different F layers arranged in an AP magnetic configuration ($\phi = 180°$). For this one purpose one can perform an analytic calculation of the ZBC in the non-self-consistent framework in situations where (as discussed in connection with Fig. 8.2) the results are not too far from those of self-consistent calculations. One finds that the ZBC at $\phi = 180°$, $G(E = 0, \phi = 180°) \equiv G_{ZB}$, for a given h and D_{F2} is:

$$G(E = 0, \phi) = \pi) = \frac{32 k_\uparrow^3 k_\downarrow^3}{A + 2\left(h^4 - 2h^2 - 2h^2 k_\uparrow k_\downarrow\right) \cos\left[2\left(k_\uparrow - k_\downarrow\right) D_{F2}\right]}.$$

(8.26)

The expression for A in Eq. (8.26) is:

$$A = a_1 \sin^2\left[(k_\uparrow + k_\downarrow) D_{F2}\right] + a_2 \left[\cos\left(2k_\uparrow D_{F2}\right)\right.$$
$$\left. - \cos\left(2k_\downarrow D_{F2}\right)\right] + a_3,$$

(8.27)

where $a_1 = 4h^2(1 - k_\uparrow k_\downarrow)^2$, $a_2 = 4h^3$, and $a_3 = h^4 + (-2 + h^2 - 2k_\uparrow k_\downarrow)^2$. I have omitted here the $\pm$ indices for the quasi-particle and quasi-hole wavevectors, since we are in the zero bias limit. In Fig. 8.4, there is a plot of the forward conductance, as given in Eq. (8.26), as a function of h for $D_{F2} = 12$ (top panel) and $D_{F2} = 18$ (bottom panel). In this zero bias limit, the circles (self-consistent numerical results) are on top of the curves (analytic results). As the thickness of the intermediate F layer increases, the number of resonance maxima and minima increases. Therefore, the resonance behavior of the ZBC is more sensitive to h for larger D_{F2}, as seen in Fig. 8.3. For a given D_{F2}, the ZBC drops considerably as ϕ varies from $\phi = 0°$ to $\phi = 180°$ when h is near the minimum of the resonance curve (rightmost panels of Fig. 8.3). On the other hand, when h is near the resonance maximum (leftmost panels of Fig. 8.3), the ZBC is a very weak function of ϕ provided that h is not too strong. By examining the denominator of Eq. (8.26), we find that the terms involving A (see Eq. (8.27)) are less important than the last term. This is because the wavelength $(k_\uparrow - k_\downarrow)^{-1}$ associated with that term is the dominant characteristic wavelength in the theory of proximity effects in FS structures, as seen in Sec. 2.7. In both panels of Fig. 8.4, we see that the ZBC for $\phi = 180°$ vanishes in the half-metallic limit. To show this analytically, one can use the conservation of probability currents and write down the following

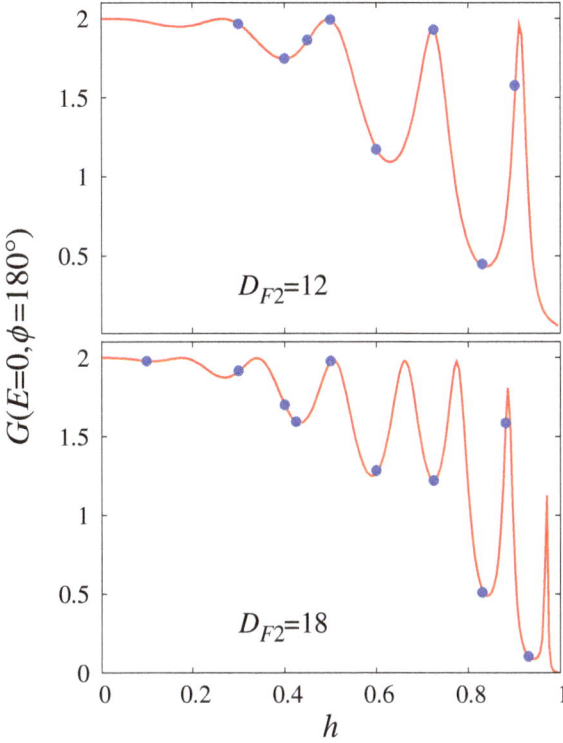

Fig. 8.4 Resonance effects in the forward scattering conductance at zero bias for trilayers at $\phi = 180°$. In the top panel, the layers have same thicknesses as in the top panels of Fig. 8.3, and in the bottom panel, they are as in the bottom panels of Fig. 8.3. The dots are the results from the computations and the curves from Eq. (8.26).

relation, valid when the bias is smaller than the bulk superconducting pair potential:

$$\frac{k_{\uparrow 1}^-}{k_{\sigma 1}^+}|a_\uparrow|^2 + \frac{k_{\downarrow 1}^-}{k_{\sigma 1}^+}|a_\downarrow|^2 + \frac{k_{\uparrow 1}^+}{k_{\sigma 1}^+}|b_\uparrow|^2 + \frac{k_{\downarrow 1}^+}{k_{\sigma 1}^+}|b_\downarrow|^2 = 1. \tag{8.28}$$

By combining Eq. (8.28) with Eq. (8.20), it becomes clear that the subgap conductances arise largely from Andreev reflection.

In the half-metallic limit, $h = 1$ in the units used in this book, conventional Andreev reflection is forbidden due to the absence of an opposite-spin band: this leads to vanishing ZBC at $\phi = 180°$. Same-spin Andreev reflection (see discussion in the paragraph above Eq. (8.1)) is not allowed in collinear magnetic configurations. Equation (8.28) also reflects another important feature of the ZBC: the contributions to G at zero bias from the spin-up and down channels are identical except for the weight factor P_σ. One can prove analytically that the sum of the first two terms (related to Andreev reflection) in Eq. (8.28) is spin-independent. As a result, the sum of last two terms, related to ordinary reflection, is also spin-independent, and so is the ZBC.

8.4.2.3 *FFS structures with magnetic layers of different materials*

It is well to consider here briefly one example where the two F layers are made of different materials, in other words, the magnitude of the exchange field in the F_1 layer is different than that in the F_2 layer. Although it is rare to find samples made this way, a brief discussion is worth having for its illustrative value. In the example below all the thicknesses and the coherence length are as in the top panels of Fig. 8.3. In Fig. 8.5, the forward scattering conductance is plotted, for several ϕ at $h_1 = 0.6$ and $h_2 = 0.1$. One can quickly spot that the ZBC here is a non-monotonic function of ϕ with it maximum at the orthogonal relative magnetization angle, $\phi = 90°$.

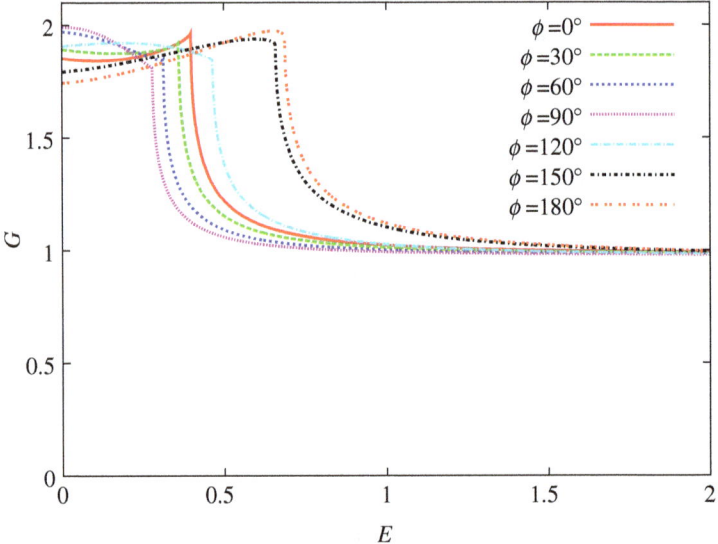

Fig. 8.5 Forward scattering conductance of a F_1F_2S trilayer with differing magnetic materials, corresponding to exchange fields of $h_1 = 0.6$ and $h_2 = 0.1$. Various magnetic orientations, ϕ, are considered as shown. Geometry and other parameters are as in the top panels of Fig. 8.3.

In contrast, results at equal exchange field strengths usually exhibit monotonic behavior, as previously shown, with the ZBC at $\phi = 0°$ being largest when the ZBC is not at its resonance maximum. However, many features are still the same, such as the formation of a FBCP (see Table 8.1) when $\phi > 90°$. For $\phi = 0°$ and $\phi = 30°$, the conductance curves are not monotonically decreasing, as was the case at $h_1 = h_2$. There, when $h_1 = h_2$ and $\phi < 90°$, there is monotonically decreasing behavior because the scattering effect due to misoriented magnetizations is not as great as at $\phi > 90°$. Also, when $h_1 \neq h_2$, we have to include in our considerations that there is an additional scattering source that comes from the mismatch between $k^{\pm}_{\uparrow 1,(\downarrow 1)}$ and $k^{\pm}_{\uparrow 2,(\downarrow 2)}$. Specifically, when $\phi = 0°$, the only important scattering effect is that due to mismatch arising from $h_1 \neq h_2$ and it leads to suppression of the ZBC at $\phi = 0°$. However, we see that the scattering due to the misoriented magnetic configuration ($\phi \neq 0°$) compensates the effect of mismatch from $h_1 \neq h_2$ and ZBC is maximized when $\phi = 90°$. Qualitatively, one can examine Eqs. (8.9) and (8.10) and verify that the spinor at $\phi = 90°$ is composed of both pure spin-up and spin-down spinors with equal weight, apart from phase factors. As a result, the scattering effect due to mismatch from $k^{\pm}_{\uparrow 1,(\downarrow 1)}$ and $k^{\pm}_{\uparrow 2,(\downarrow 2)}$ is reduced. One can also verify that, when the strength of h_2 is increased towards h_1, the locations for the maximum of the ZBC(ϕ) curves gradually move from $\phi = 90°$ at $h_2 = 0.1$ to $\phi = 0°$ at $h_2 = 0.6$.

8.4.3 *Angularly averaged results and anomalous (equal spin) Andreev reflection*

We will now discuss some results for the angularly averaged conductance, $\langle G \rangle$ as defined in Eq. (8.21). The angularly averaged conductance is relevant to a range of experimental results where consideration of the forward conductance alone, which is measured in point contact experiments, is insufficient. The details of the angular averaging are explained under Eq. (8.23). The averages plotted in this section are over all angles up to the lesser of $\pi/2$ or the critical angle $\theta_{c\sigma}$ as discussed below Eq. (8.23). In general, however, the maximum angle of integration and the weight factor for angular averaging in Eq. (8.21) are to be determined based on a real experimental set-up or on the geometry of the junction: sometimes neither the forward nor the "all angles" limits may be appropriate. Physically, considerations based on the angularly averaged results given here will naturally lead us to the introduction of another phenomenon: the *equal spin* or *anomalous* Andreev reflection.

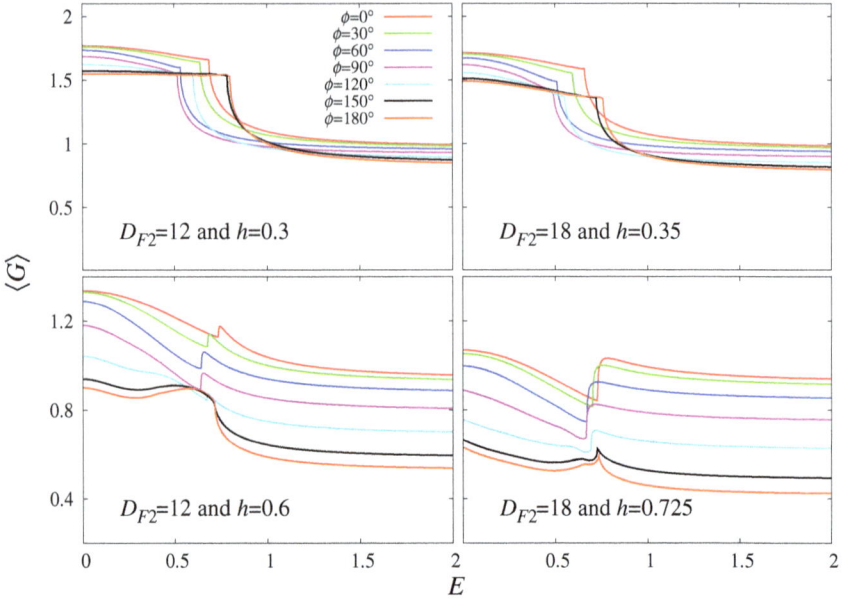

Fig. 8.6 Bias dependence of the angularly averaged conductance of $F_1 F_2 S$ trilayers for several angles ϕ (see legend). In the left panels, $D_{F1} = 10$, and $D_S = 180 = 1.5\Xi_0$, as in the top panels of Fig. 8.3. In the right panels, $D_{F1} = 10$ as in the bottom panels of Fig. 8.3. In all cases, the ZBC decreases with increasing ϕ.

In Fig. 8.6, we can see some results for $\langle G \rangle$ at $D_{F2} = 12$ (left panels) and $D_{F2} = 18$ (right panels). All curves are for $D_{F1} = 10$ and $D_S = 180 = 1.5\Xi_0$. The values of h are indicated in each panel. Results are shown over the entire range of ϕ values. The CB values obtained for $\langle G \rangle$ are again non-monotonic functions of ϕ and the non-monotonicity matches that of the saturated pair amplitudes, for the reasons previously given. The CB values for $\langle G \rangle$ in these cases are the same as those for the forward scattering conductance. One can also see that the resonance phenomena tend to get washed out in the angularly averaged conductance. For example, the resonance curve in the top panel of Fig. 8.5 tells us that $h \approx 0.3$ and $h \approx 0.6$ correspond respectively to a resonance maximum and minimum of the ZBC in the forward scattering G. However, in the top left panel of Fig. 8.6, the ZBC is no longer a weak function of ϕ and it gradually decreases when ϕ is increased. Near the resonance minimum, $h = 0.6$ (bottom left panel of Fig. 8.6) we can see a trace of the appearance of the FBCP (see Table 8.1

for acronyms) when ϕ is above $90°$. This FBCP in $\langle G \rangle$ is not as prominent as that in the forward scattering G, due to the averaging.

The magnetoresistance measure $M_G(E = 2, \phi)$ (see Eq. (8.25))is larger for $\langle G \rangle$ than for the forward scattering conductance. For example, $M_G(E = 2, \phi = 180°)$ in the forward scattering conductance for $h = 0.6$ and $D_{F2} = 12$ is half of that in $\langle G \rangle$. As for the zero bias magnetoresistance $M_G(E = 0, \phi = 180°)$ in $\langle G \rangle$, it is of about the same order as $M_G(E = 2, \phi = 180°)$ and it does not depend on where it is located in the resonance curve shown in Fig. 8.5: recall that $M_G(E = 0, \phi = 180°)$ for the forward scattering conductance almost vanishes at the resonance maximum.

In the right-side panels of Fig. 8.6, results are plotted for a larger D_{F2} with values of $h = 0.35$ (near a resonance maximum) and $h = 0.725$ (near a resonance minimum). They share very similar features with the thinner D_{F2} case in the left panels. However, for $h = 0.725$, the ZBC values at different ϕ shrink to almost or less than unity and they are just barely higher than the conductance at $E = 2$ because the contributions from Andreev reflection are strongly suppressed in such a high exchange field.

Another prominent feature in the angularly averaged results for higher exchange fields (bottom panels in Fig. 8.6) is the existence of cusps at the CB. To understand the formation of these cusps, it is useful to analyze $\langle G \rangle$ by dividing the contribution from all angles into two parts, the range above and the range below the conventional Andreev critical angle θ_c^A (recall the discussion below Eq. (8.24)). Consider first the case of spin-up incident quasi-particles. We see that when $\theta_c^A \equiv \sin^{-1}[\sqrt{(1 - h)/(1 + h)}] < \theta_i < \sin^{-1}[\sqrt{1/(1 + h)}]$, the conventional Andreev reflected waves become evanescent while the transmitted waves are still traveling waves above the CB. When $\theta_i > \sin^{-1}[\sqrt{1/(1 + h)}]$, both the conventional Andreev reflected waves and the transmitted waves become evanescent. In this case, $\theta_{c\uparrow} = \sin^{-1}[\sqrt{1/(1 + h)}]$ is the upper limit in Eq. (8.21): the angular integration limits for the spin-up component of $\langle G \rangle$ are nontrivial. On the other hand, for the case of spin-down incident quasi-particles the limits are trivial, because the dimensionless incident momentum is $\sqrt{1 - h}$ which is less than both the conventional Andreev reflected wavevector, $\sqrt{1 + h}$, and the transmitted wavevector, (unity in our conventions). Therefore, all the reflected and transmitted waves at $E \approx 1$ are traveling waves. As a result, all possible incident angles are included and the upper limit of Eq. (8.21) is $\pi/2$.

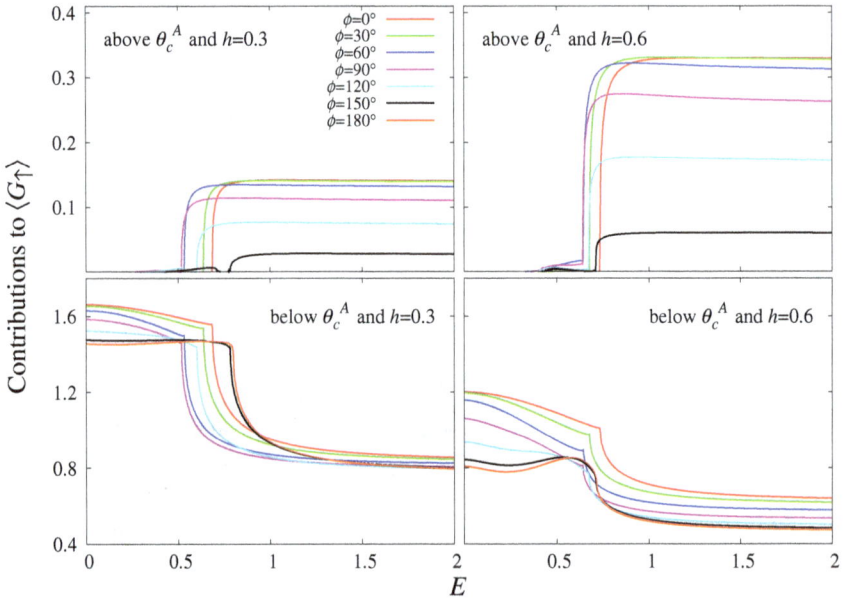

Fig. 8.7 Contributions (see text) to the spin-up angularly averaged conductance, $\langle G_\uparrow \rangle$, from angular ranges above (top panels) and below (bottom panels) the Andreev critical angle θ_c^A. Values of ϕ considered are indicated. The top panel results at $\phi = 180°$ are vanishingly small. The geometric and exchange field values are as in the left panels of Fig. 8.6.

To illustrate the consequences of this, the contributions to $\langle G_\uparrow \rangle$ from angles in the range above θ_c^A (top panels) and below that angle (bottom panels) are separately plotted in Fig. 8.7. This is done for the field values and geometry in the left panels of Fig. 8.6, in particular $D_{F2} = 12$. These contributions are denoted here as $\langle G_\uparrow(E) \rangle_{above}$ and $\langle G_\uparrow(E) \rangle_{below}$ respectively. The $\langle G_\uparrow(E) \rangle_{below}$ contributions, in the bottom panels of Fig. 8.7, are for both $h = 0.3$ and $h = 0.6$ similar to the result for their total forward scattering counterpart (see Fig. 8.2 and the upper right panel of Fig. 8.3). Of course, the angular averaging leads to some smearing of the pronounced features originally in the forward scattering G. Qualitatively, the similarity comes from the propagating nature of all possible waves except the transmitted waves below the CB when $\theta_i < \theta_c^A$. Therefore, the forward scattering G is just a special example with the incident angle perpendicular to the interface.

In the subgap region, the contribution to $\langle G_\uparrow(E)\rangle_{above}$ is vanishingly small although small bumps appear when the exchange fields in the two F layers are non-collinear, i.e., $\phi \neq 0, \pi$. These small bumps are generated by a process which is called "anomalous Andreev reflection" (or "equal spin AR") . This process differs from the standard Andreev reflection in that the reflected hole has the *same* spin as the incident particle. This involves singlet to triplet pair conversion and it can occur under the same circumstances as the generation of $m = \pm 1$ triplet pairing correlations becomes possible. Hence, triplet generation is correlated with this anomalous Andreev reflection. In other words, equal-spin Andreev reflection is possible in trilayers because, in a non-collinear magnetic configuration, a spin up quasiparticle can Andreev reflect as a spin-up hole. This can be seen from the matrix form of the BdG equations, Eq. (2.8). The occurrence of anomalous Andreev reflection leads to some important physics which we shall further discuss below. One can see from Fig. 8.7, that when the exchange fields are strictly parallel or anti-parallel to each other, equal spin Andreev reflection does not arise.

Above θ_c^A, the conventional Andreev-reflected wave is evanescent and it does not contribute to $\langle G_\uparrow\rangle$. When the bias is above the saturated pair amplitude, contributions to $\langle G_\uparrow\rangle$ from the upper range are provided by both the transmitted waves and by anomalous Andreev reflected waves: recall that ordinary transmitted waves are propagating when E is greater than the saturated pair amplitudes. One can also see on the figure that $\langle G_\uparrow\rangle_{above}$ decreases with increasing ϕ. At $\phi = 180°$, $\langle G_\uparrow\rangle_{abpve}$ is vanishingly small due to the effect of a large mismatch from the anti-parallel exchange field. Also, the contribution from above θ_c^A is seen to be less in the $h = 0.3$ case than at $h = 0.6$. This is mainly due the fraction of states with incident angles larger than θ_c^A being smaller at $h = 0.3$. On the other hand, the contribution from below θ_c^A is larger in the $h = 0.3$ case. The increase of $\langle G_\uparrow\rangle_{above}$ and the decrease of $\langle G_\uparrow\rangle_{below}$ from $h = 0.3$ to $h = 0.6$ gives rise to the cusp at the CB, when adding these two contributions together.

Equal-spin (anomalous) Andreev reflection (ESAR or AAR), which can be generated when the magnetic configuration is non-collinear, is worth further discussion, particularly because of its connection to triplet generation. We have seen (recall the discussion below Eq. (8.24)) that conventional Andreev reflection is forbidden when $\theta_i > \theta_c^A = \sin^{-1}(\sqrt{(1-h_1)/(1+h_1)})$. Thus, θ_c^A vanishes in the half-metallic limit, $h = 1$. Then conventional Andreev reflection is not allowed for any incident angle θ_i and the subgap

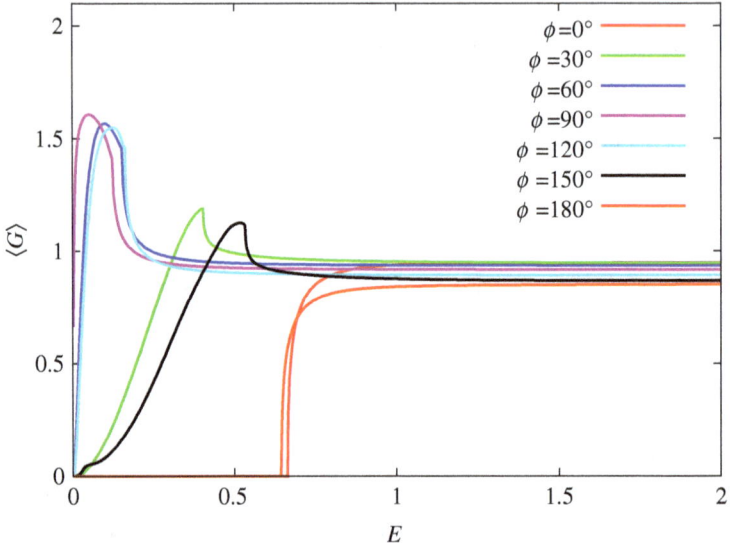

Fig. 8.8 The angularly averaged conductance of F_1F_2S trilayers with exchange field $h_1 = 1$ and $h_2 = 0.1$ for several values of ϕ. At small but nonzero E the curves from higher to lower are for $\phi = 90°, 60°, 120°, 30°, 120°$, with those corresponding to zero or $180°$ being zero, in the subgap region, as triplets cannot be generated and hence there is no ESAR. See text for discussion.

$\langle G_\uparrow \rangle$ arises only from ESAR. To illustrate this we now show, in Fig. 8.9, results for a trilayer structure consisting of a half-metal ($h_1 = 1$) and a much weaker ($h_2 = 0.1$) ferromagnet. The weaker ferromagnet serves the purpose of generating ESAR. It is also possible to generate ESAR in metal-superconductor bilayers with spin-flip interface. This interface plays the same role as the weaker ferromagnet here. In the trilayers discussed here the angle between internal fields must be $\phi \neq 0, \pi$.

In Fig. 8.8, are plotted results for the $\langle G \rangle$ of this particular system for several ϕ. The geometrical parameters are again $D_{F1} = 10$, $D_{F2} = 12$, and $D_S = 180$. We have $\langle G \rangle = \langle G_\uparrow \rangle$ because the weight factor $P_\downarrow = 0$ in this half metallic case. For $\phi = 0°$ and $\phi = 180°$ the CB value is about 0.65 and, below the CB (in the subgap region), $\langle G \rangle$ vanishes because the conventional Andreev reflection is completely suppressed and ESAR is not allowed in the collinear cases. For $\phi = 30°$ and $\phi = 150°$, the CB is near 0.4 and 0.5 respectively and all of the subgap $\langle G \rangle$ is due to ESAR. The CB values for $\phi = 60°$, $90°$, and $120°$ are 0.15, 0.12, and 0.15. For these

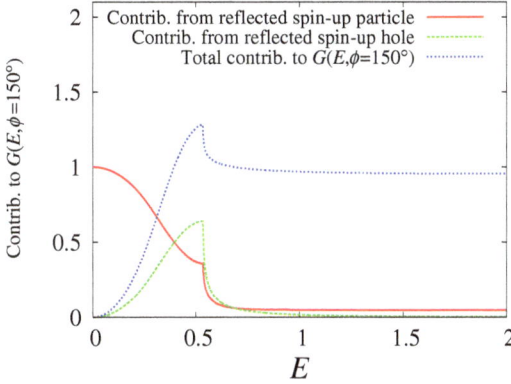

Fig. 8.9 Contributions to $G(E, \phi = 150°)$, computed for the parameter values used in Fig. 8.8, from the spin-up quasiparticle and spin-up quasihole ESAR (see text for discussion). The total G is also shown.

three angles, a FBCP clearly forms, arising from the ESAR in the subgap region.

The conductance in the subgap region, which is in this case due only to ESAR, can be isolated by choosing the $\phi = 150°$ angle and plotting, in Fig. 8.9, the contributions to G (for this case G and $\langle G \rangle$ are very similar) from the reflected spin-up particle and the reflected spin-up hole wavefunctions. The spin-down particle and spin-down hole wavefunctions are evanescent and do not contribute to the conductance. Thus, Eq. (8.20) becomes

$$G = 1 + (k_{\uparrow 1}^- / k_{\uparrow 1}^+)|a_\uparrow|^2 - |b_\uparrow|^2. \tag{8.29}$$

The quantities shown are the second (solid curve labeled "from spin up hole") and third (highest curve at the origin) terms in this expression. The value of G is also plotted. Below the CB the relation for conservation of probability current, $[k_{\uparrow 1}^- / k_{\sigma 1}^+]|a_\uparrow|^2 + [k_{\uparrow 1}^+ / k_{\sigma 1}^+]|b_\uparrow|^2 = 1$, is satisfied. One sees that the reflected ESAR amplitudes decay very quickly above the CB. However, these reflected amplitudes are quite appreciable in the subgap region. In other words, the supercurrent in the subgap region contains signatures of the triplet correlations. This confirms the simple picture [25] that above the CB the current flowing throughout the junction is governed by the transport of quasiparticles. However, below the CB it is dominated by ESAR.

8.4.4 *The conductance of spin valve structures: General*

Now we turn to the spin valve geometry of Fig. 2.1. As explained in Chap. 6, the presence of an N layer separating the two ferromagnetic ones is absolutely necessary from the experimental point of view: therefore the results in this subsection are, from that point of view, much more important than those in the previous one. An important aspect that I will emphasize is the reasons why these "spin valve structures" have very unusual properties as spin valves. This will be discussed here in connection with the results.

The self-consistent pair potential $\Delta(y)$ is again extracted as explained in Sec. 8.3.1 and Chap. 2. The more complicated geometry introduces primarily matters of notation. The main one arises because we have now three interfacial barriers to consider: we do want very much to study in some detail the effects of unavoidable scattering roughness. It is convenient to number these interfaces as going from right to left in Fig. 2.1 so that the interfacial potential Eq. (2.10) is

$$U(y) = H_1\delta(y - d_{f1}) + H_2\delta(y - d_{f1} - d_N) + H_3\delta(y - d_{f1} - d_N - d_{f2}).$$
$$(8.30)$$

I will, as before, use the dimensionless parameters $H_{Bi} \equiv H_i/v_F$, where v_F is the Fermi speed in S, to conveniently characterize the strength of the delta functions.

As to the transfer matrix methods, they are exactly as in Sec. 8.3.2 and I recall here that the wavefunctions in the N layer are of the same form as in the intermediate magnetic layer, Eq. (8.10), but with the internal field set to zero, which simplifies matters very much.

The spin valve structure discussed in this subsection is of course the same whose thermodynamic properties were studied, and successfully compared with experimental results in Chap. 6. It therefore makes sense to focus in the parameter ranges that were shown in that chapter to best describe actually fabricable samples. This applies in particular to the dimensionless barrier strengths, H_{Bi} for which one should use intermediate values appropriate to "good", not perfect, samples. Values of these parameters close to or higher than unity would represent a strong tunneling limit: these would be experimentally very undesirable as the proximity effects would be very small. Zero values represent an ideal interface, which is unattainable experimentally. Since the first and second interfaces are both between F and N materials, one can fairly safely assume that these two barrier strengths are similar, and in the results we will show they are usually taken to be identical, $H_{B1} = H_{B2} \equiv H_B$.

As to the internal field strength parameters h_1, h_2 for F_1 and F_2 respectively, it is safe to assume that they are the same, since in most experiments (see Chap. 6) the same magnetic material is employed, for good reason. The results for G presented are for $h = 0.145$, a value which we have previously found adequate in Chap. 6 to fit Co related properties in these devices. Also as in Chap. 6 we show results here for $\Lambda = 1$, which subsumes the scattering effect due to wavevector mismatch with the phenomenological H_{Bi} parameters. Similarly, the results are obtained for the value of $\Xi_0 = 115$ of the dimensionless correlation length in S, the value used in the same context (again, see Chap. 6) for Nb. The thicknesses of all layers are also varied in this section within experimentally relevant ranges, while not keeping solely to the range used in Chap. 6. The thickness of the inner ferromagnetic layer, D_{F2}, will be taken relatively small, which is necessary to obtain good proximity effect, while allowing D_N and D_{F1} to be somewhat larger. As to D_S, the thickness of the superconducting layer, it must of course be kept above Ξ_0: otherwise the sample tends to become non-superconducting, for obvious reasons. In this subsection I will focus on forward conductance results, which can be obtained from point probes and involve trends easier to understand. We have discussed the differences between forward and angularly averaged results quite extensively in the previous subsection and it would be tedious to repeat the exercise, which leads to the same physical conclusions anyway.

8.4.5 *Spin valve conductance: Effects of the interfacial scattering*

The effects of interfacial scattering are very strong and important and we discuss them first and carefully. Even in standard normal-superconductor interfaces the zero bias conductance (ZBC) can vary between a value of two for a perfect interface, and an exponentially small value for the tunneling limit, as explained in Sec. 7.1.2. One should recall also that it is impossible for the two Fermi wavevectors in the ferromagnets to match the Fermi wavevector of either the N or the S materials. This has to be kept in mind in the discussion below.

Let us begin by showing in Fig. 8.10 the effect of increasing H_{B3} (see Eq. (8.30) and the paragraph that follows it) only, assuming for the time being that the other interfaces are perfect. Four values of H_{B3} are studied, one in each panel, and curves for seven values of the misalignment angle ϕ, which is a very important variable here, are plotted. The geometrical parameters are $D_{F1} = 20$, $D_N = 40$, $D_{F2} = 12$ and $D_S = 180$. The overall

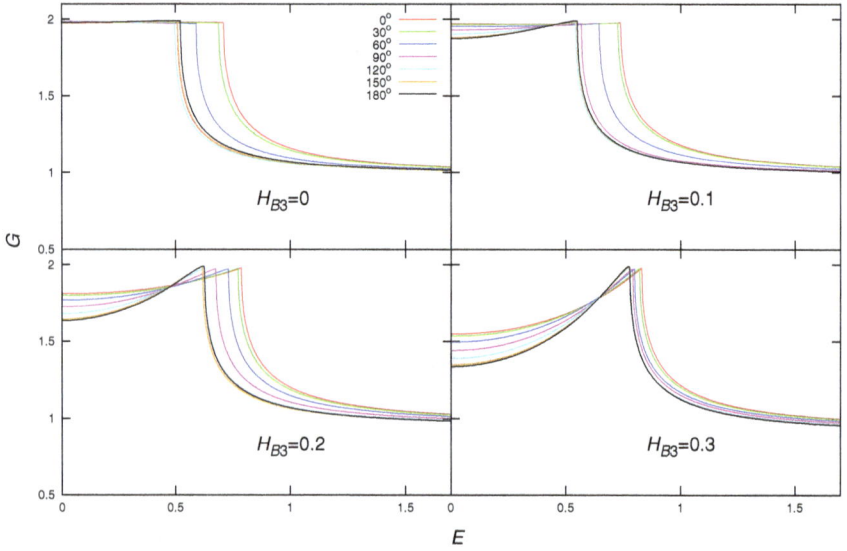

Fig. 8.10   The bias dependence of the forward conductance of a spin valve $F_1 N F_2 S$ structure (Fig. 2.1). The effect of scattering at the interface between S and F_2 is emphasized, as characterized by H_{B3}. Results are for seven values of the misalignment angle ϕ (legend). The values of H_{B3} are indicated, and $H_{B1} = H_{B2} \equiv H_B = 0$. The thicknesses are $D_{F1} = 20$, $D_N = 40$, $D_{F2} = 12$ and $D_S = 180$. The internal field parameter is $h = 0.145$

trend on increasing H_{B3} is a marked decrease of the low bias conductance and a much smaller decrease of the high bias limiting value. The critical bias (CB), which, as we recall, is the value of the bias at which G clearly changes behavior and begins trending towards its normal state limit, is in general smaller than unity. Smaller values reflect stronger proximity effects since the CB is associated with the saturated value of $\Delta(Y)$ well inside S. One can see accordingly that the CB tends to increase with H_{B3}, while the value of G at critical bias (the critical bias conductance, CBC, see Table 8.1) remains nearly the same. On the other hand, the CB is in all cases a strong function of ϕ, decreasing as ϕ increases, up to about $\phi = 100°$ and then flattening, for this geometry. The ϕ dependence is less marked at higher barrier values, when the proximity effects are weaker. The ZBC, however, is basically monotonically decreasing in ϕ. This dependence on ϕ is different from that of the CB or CBC, and it leads to a crossover in the conductance values. Remarkably, this crossover tends to occur with a "nodal" behavior at a single bias value in the subgap region: this can best be seen in

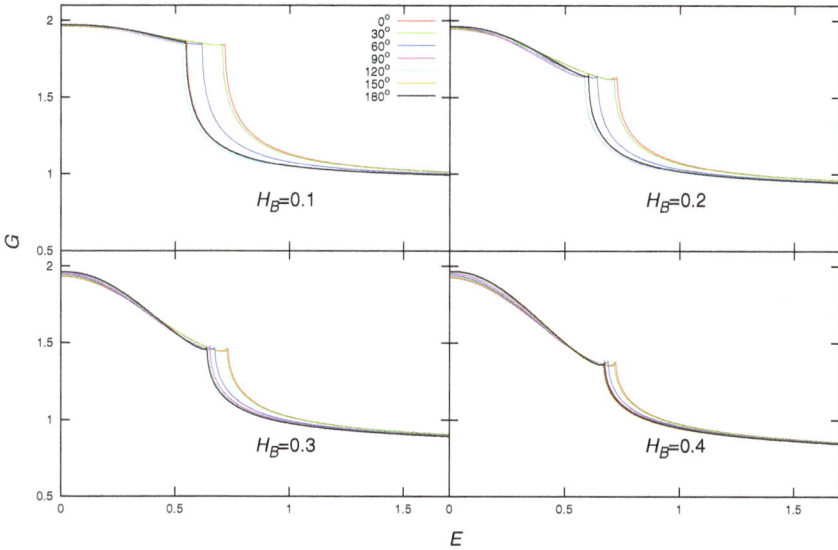

Fig. 8.11   Effect on G of the barriers between N and the ferromagnets in a $F_1 N F_2 S$ structure. Here $H_{B1} = H_{B2} = H_B$ is varied while $H_{B3} = 0$. The four panels show results for the same arrangement as in Fig. 8.10 and the same geometrical and field parameters.

the third and fourth panels of Fig. 8.10. Monotonic behavior in the ZBC also occurs for other values [30] of D_{F2} but the direction (increasing or decreasing in ϕ) is reversed in an oscillatory way: for instance the ZBC increases with ϕ at values of D_{F2} of 7 and 10 and again at 16,17. This is one more example of the multiple oscillatory behavior found in this problem and an illustration of how much care one has to take before extrapolating results.

Next, in Fig. 8.11, we see the effect of increasing $H_{B1} = H_{B2} \equiv H_B$ (see again Eq. (8.30) and the discussion below it) while keeping $H_{B3} = 0$ at the F_2/S interface. Again, four barrier values are considered, in an arrangement very similar to that in Fig. 8.10. The effects of interfacial scattering are now more pronounced. This is not necessarily due to the presence of two barriers instead of one. As in well known situations in elementary one-dimensional quantum mechanics, having more barriers does not necessarily lead to less transparency: there are resonance phenomena involved. This is also the case here, although the analogy is imperfect: our system is not one-dimensional; there are multiple scattering mechanisms (interfacial imperfections, wavevector mismatch, Andreev reflection, etc.). Still, having two barriers does not always reduce transmission. A clear example of this can be seen in the ZBC value which, for the chosen values of $D_{F2} = 12$ and D_N, is nearly independent of H_B. This is because of resonance-like behavior in this geometry. Furthermore, changing the values of $D_{F2} = 12$ and D_N leads to ZBC behavior more similar to that in Fig. 8.10, which is discussed in the next subsection in connection with Fig. 8.14. The behavior of the CB with angle is nonmonotonic, in a way similar to that seen in Fig. 8.10. The minimum is now somewhat less shallow, particularly at higher H_B. At low bias, G decreases as the bias is increased, although an upturn does occur as the CB is approached albeit at a lower value of the CBC for increasing H_B. This is in contrast to Fig. 8.10 where the CBC was unaffected by H_{B3}.

8.4.6   *Spin valve conductance: Effects of geometry*

It is rather obvious at this point that in addition to the barrier strengths the geometry of the spin valve will influence the conductance and should be studied. Indeed, we have just seen in the previous discussion a hint that the thickness of the different layers may have a strong and often nonmonotonic effect on G. The thickness of the inner magnetic layer, D_{F2}, turns out to be the more important of these geometrical variables. In the six panels in Fig. 8.12 we see results for G, at several values of the misalignment angle ϕ, at increasing values of D_{F2} while the other geometrical parameters are kept constant, as are the material parameters, all fixed to their values in the previous two figures. The three interfacial barrier parameters H_{Bi} are set to 0.3, 0.3, and 0.1 for $i = 1, 2, 3$ respectively (see Eq. (8.30)). These are intermediate and experimentally realistic values.

Consider in detail the first panel, where $D_{F2} = 7$. One notices immediately the reduction in ZBC, as opposed to the results for $D_{F2} = 12$ in the third panel or to those in the previous figures. The dependence of this reduction on D_{F2} is, as has been mentioned above, oscillatory: after disappearing in the second and third panels, it can be seen again at $D_{F2} = 15$ (fourth panel). In this first panel, as in the second and the fifth, the minimum value of the CB as the angle ϕ varies is near $\phi = 90°$, and this minimum is very well marked. This is an optimum situation for valve effects: G depends strongly on ϕ so one can exhibit the valve effect by varying this angle, using methods such as those discussed

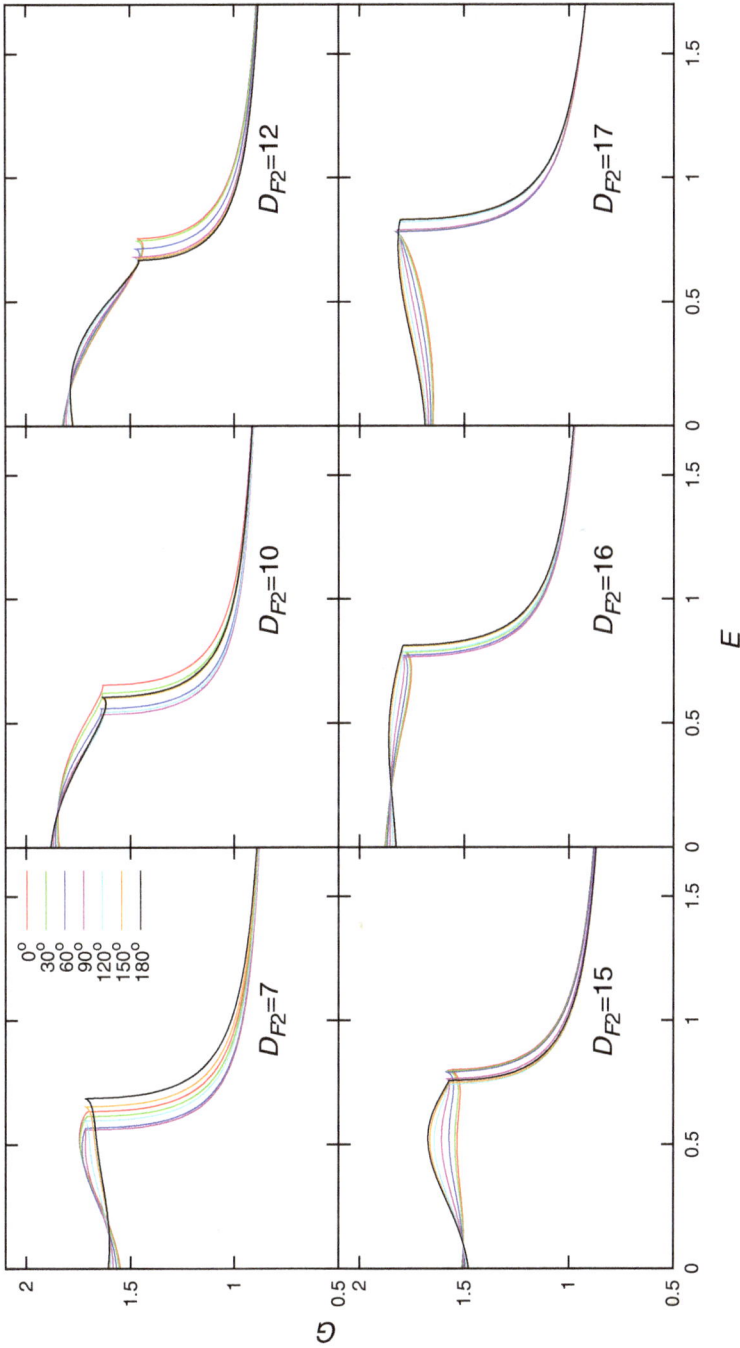

Fig. 8.12 Effect on G in a F_1NF_2S structure of varying the thickness D_{F2} of the inner F layer. The values of the other thicknesses, field, and correlation length are as in the previous two figures, and the barrier heights are set to experimentally realistic values (see text). The panels show G vs. E for several values of ϕ, at six values of $D_{F2} = 7, 10, 12, 15, 16,$ and 17. The spin valve effect varies significantly in both the CB and the ZBC.

in Chap. 6. The ZBC value depends somewhat on ϕ but not in the same way as the CB: hence, the conductance curves cross, forming a node near a bias of 0.2. The second panel exhibits similar behavior, but the ZBC is markedly higher. On further increasing D_{F2} to 12 (third panel) the CB becomes monotonic in ϕ while the low bias conductance does not change: indeed the node where the lines cross barely changes location between the second and third panels. The case $D_{F2} = 15$ (fourth panel) is yet different: the CB is larger and there is a marked "bump" in the low bias conductance, the height of which increases with ϕ. The oscillatory effects in the ZBC are observed again in its increase in the fifth panel — there the angular dependence of the CB returns to having a marked minimum at $\phi = 90°$ although with a weaker dependence. Furthermore, the node noticeably moves to a higher bias value. Finally, at $D_{F2} = 17$ (last panel) the ZBC drops again, the angular dependence of the CB is reversed, and the node disappears. Thus we see that the thickness of the inner magnetic layer is a very important variable in determining the conductance properties.

One should examine next the effect on G of varying the thicknesses of the outer F layer and the N layer in the valve. This is done in Fig. 8.13 where there are plotted, in each panel, results for G at fixed $\phi = 0$.

The first thing to observe is that the effect of varying D_{F1}, the thickness of the outer ferromagnetic layer, is much weaker than that of varying D_{F2}. This is illustrated in the first two panels of Fig. 8.13. In the first one results for $G(E)$ are plotted for several values of D_{F1} ranging from 12 to 30 and, in the second panel, for D_{F2} values from 7 to 17 at fixed D_{F1}. In both panels $D_N = 40$. Barrier heights and other parameters are precisely as in Fig. 8.12. The difference is obvious: while in the first panel the results barely change (although the change is nonmonotonic), in the second one every relevant quantity (CB, ZBC, high bias and low bias behaviors etc.) changes, in obvious and very strongly nonmonotonic ways. Thus, in the fabrication process, the precise thickness of D_{F1} is much less critical than that of D_{F2}. As to the normal spacer thickness, in the last two panels of Fig. 8.13 the dependence of G on D_N is examined. Again, G is plotted at fixed $\phi = 0$ for several values of D_N at two values of D_{F2} (see caption). One can see that while quantities such as the CB do not depend very much on D_N, the low and high bias behaviors vary quite appreciably overall, the former rather dramatically. Hence one can conclude that D_{F2} is the crucial geometrical parameter in the problem, followed in importance by D_N and with D_{F1} being much less relevant.

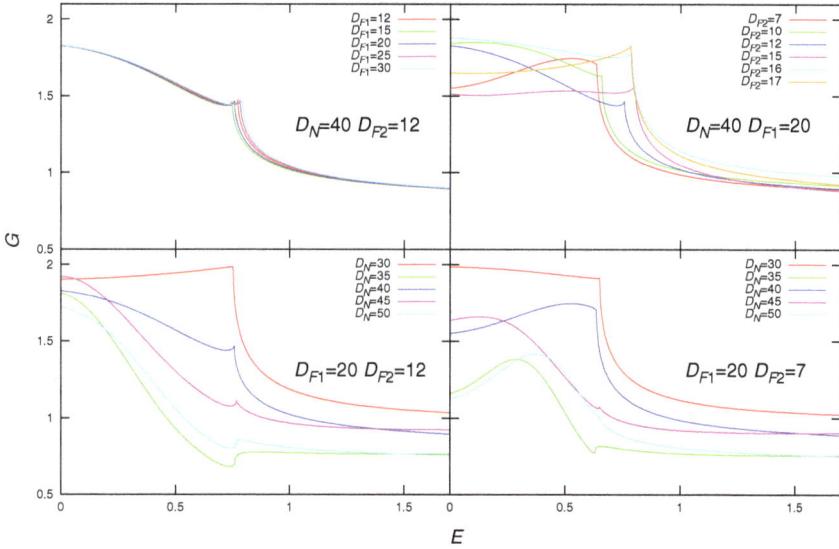

Fig. 8.13 Effects on the forward conductance of varying D_{F1} or D_N, in the spin valve, contrasted with the dependence on D_{F2}, Fig. 8.12. All panels are for $\phi = 0$. Barrier values, the field parameter, correlation length, and D_S are as in Fig. 8.12. The first two panels contrast the effect on G of varying D_{F1} with that of varying D_{F2}. In the first panel, D_{F1} is varied, as indicated in the legend, at $D_{F2} = 12$, while in the second one D_{F2} is varied at $D_{F1} = 20$. The third and fourth panels show the effect of varying D_N at $D_{F1} = 12$ and $D_{F2} = 7$ respectively. The dependence of the results on D_{F1} is much weaker than that on D_{F2} or D_N. Both D_{F2} and D_N have a large impact on the ZBC, while D_{F2} has a much larger effect on the CB.

Careful examination of the above results yields insights on the combined effects of interfacial scattering and of geometry, for the latter particularly on the effect of D_{F2}. How the effects of geometry and interfacial strength are related follows ultimately the oscillatory nature of the Cooper pair amplitudes and from quantum mechanical interference. In Fig. 8.14, these combined effects are exhibited in a more direct way. As in Fig. 8.13 the plotted results are for fixed $\phi = 0$. There are four values of D_{F2} considered, one in each panel, ranging from $D_{F2} = 7$ to $D_{F2} = 17$, and results are shown for several values of H_B at $H_{B3} = 0$. In the first panel one sees a large and monotonic dependence on H_B of the entire conductance behavior. In the next case shown, $D_{F2} = 12$, the ZBC depends only very weakly on H_B. In the next one, the spread in the ZBC with ϕ increases somewhat, as compared to the previous panel, and it does so even more in the last panel. This resonance-like behavior is not the precise analog of the situation in the

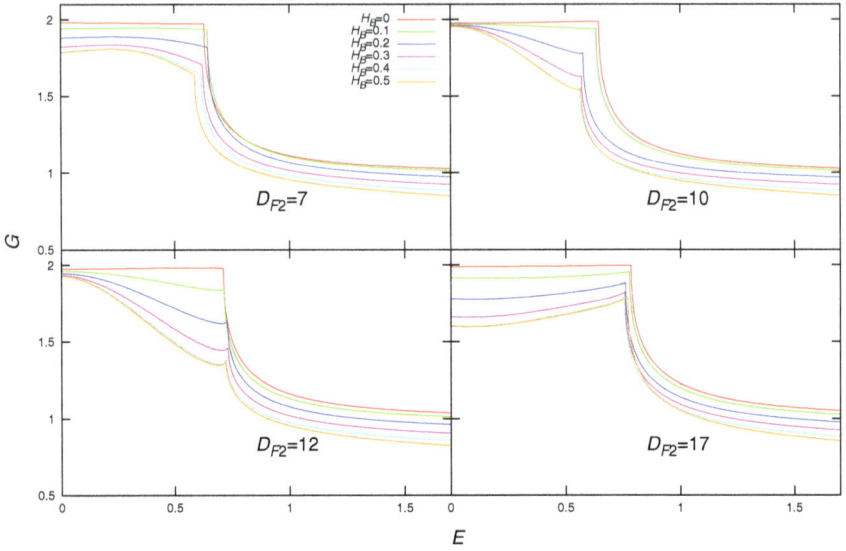

Fig. 8.14   Combined effect on $G(E)$ of D_{F2} and interfacial barriers. The behavior at fixed $\phi = 0$ and $H_{B3} = 0$ is studied. Each of the four panels corresponds to a fixed value of D_{F2}: 7, 10, 12, and 17 and the curves correspond to values of $H_{B1} = H_{B2} \equiv H_B$ as indicated in the legend. A nonmonotonic feature in the ZBC is observed as a function of D_{F2}, owing to the oscillatory behavior of the Cooper pairs.

one-dimensional two barrier problems in basic quantum mechanics, where a resonance feature exists in the transmission coefficients as a function of the distance between the barriers. This analogy might apply better to D_N, but not to the inner ferromagnetic thickness D_{F2}. Instead, this resonance is primarily due to the oscillatory behavior of the Cooper pairs. We see then that certain values of D_{F2} make the system, or at least its ZBC, partly "immune" to the effects of fairly high surface barriers. Although this holds only to a limited extent, it may be worthwhile to attempt to exploit this effect to palliate the existence of unfavorable interfaces with unavoidably large scattering.

8.4.7   *Temperature dependence*

Here is a good place to briefly discuss the temperature dependence of the results given, a subject often omitted in the literature. One can not always assume that the experiments are performed in the low T limit, although this is usually attempted.

At finite temperature there are two obvious sources of corrections to the $T \to 0$ results. The first and more obvious is that arising from the T dependence of $\Delta(y)$, that is, the T dependence of the effective BCS Hamiltonian. This dependence is of course straightforward to include: one just calculates the self-consistent Δ at finite T via Eq. (2.12), which has explicit T dependence and uses it as input in the transfer matrix calculations. There are also the usual temperature corrections of order T/T_F which one finds in all solid state calculations. These are of course negligible.

But there is a also a less obvious source of temperature dependence, of leading order $(T/\Delta_0)^2$. This arises from corrections to Eq. (8.20) itself. That expression, derived from simple quantum mechanical arguments (see the discussion in connection with Eq. (7.2)) is valid only at $T = 0$. The same physical arguments given in Secs. 7.1.1 and 7.1.2, combined with Ohm's law, give the more general expression for the current:

$$I(V) = \int G_0(\epsilon) \left[f\left(\epsilon - eV \right) - f\left(\epsilon \right) \right] d\epsilon, \tag{8.31}$$

where f is the Fermi function. The bias dependent tunneling conductance is, at any temperature $G(V) = \partial I / \partial V$. The function G_0 in Eq. (8.31) is the conductance, as given by Eq. (8.20), but with the self-consistent gap calculated at finite temperature. This can easily be seen by replacing the temperature derivative of the Fermi function by a δ function. The two terms in the integral on the right side of Eq. (8.31) have obvious physical meaning that show the correctness of the expression. As before, one can look for a less intuitive but more rigorous derivation in Ref. [25].

So, there is a temperature dependence arising from the Fermi function in Eq. (8.31). If the temperature is not too close to T_c^0, the transition temperature of the bulk S material, which sets the overall scale, one can use a Sommerfeld type expansion. Because the energy scale over which $G(V)$ varies is of order Δ_0, as we have already repeatedly seen, the relevant expansion parameter is T/T_c^0, not T/T_F, and hence not necessarily negligibly small in all experimental situations. One finds using textbook [31] Sommerfeld expansion methods:

$$G(V,T) = G_0(V) + a_1 \left(\frac{T}{\Delta_0} \right)^2 \left(\frac{\partial^2 G(V)}{\partial \epsilon^2} \right)\Big|_{\epsilon=V} + \mathcal{O}\left(\frac{T}{\Delta_0} \right)^4 \tag{8.32}$$

where a_1 can be expressed [31] in terms of a Bernoulli number.

Alternatively, one can use the general form:

$$G(V,T) = \frac{1}{4T} \int dV' \frac{1}{\cosh^2[(1/2T)(V-V')]} G_0(V'). \qquad (8.33)$$

In Eqs. (8.32) and (8.33) $G_0(V)$ means the result of Eq. (8.20) evaluated with the self-consistent pair potential at temperature T. The second of the above expressions, Eq. (8.33), turns out to be more useful in practice, particularly if relevant temperatures turn out to be too high for the Sommerfeld expansion. An entirely similar argument starting with Eq. (8.21) shows how the above expressions can be combined with angular averaging.

Recapitulating, we see that there are two relevant sources of T dependence. The first is that arising from the self-consistent pair potential, $\Delta(y)$, (Eq. (8.11)) that is, the T dependence in the effective Hamiltonian. This leads to the function G_0 defined below Eq. (8.31) and in Eq. (8.20) being implicitly T dependent. This effect is important near the transition temperature, but not otherwise. The second source is that originating in the Fermi functions in Eq. (8.31). In practice, Eq. (8.33), which is not dependent on any expansion, is much more useful than the Sommerfeld method in most relevant temperature ranges. This is, besides the reasons already mentioned, because the conductance has large, and even discontinuous derivatives with respect to voltage, which the Sommerfeld expansion does not handle well.

Some representative results are shown in Fig. 8.15. In the top two panels the misalignment angle is fixed, $\phi = 0$, and results are shown for G both at $T = 0$ and at a reduced temperature $t = 0.1$, (recall Table 2.1). Since for the size ranges considered in this section the T_c/T_c^0 values are in the 0.5 to 0.6 region, these correspond to T/T_c of about 0.2. The first panel shows results in a strong tunneling limit regime, with high barriers, and the second for zero barrier heights. Plots of G_0 , i.e., the results obtained by using the $\Delta(y)$ correction only, are also included: these are obviously inadequate in both cases, and the full result is needed. This is invariably the case except at extremely low T: the correction from the Fermi functions is dominant. The overall effect of the temperature is, otherwise, that of rounding up and softening the sharp features of the low T results. A consequence of this is that at finite T one has to redefine more carefully the CB as the bias value at which G has a peak or a high derivative. The proper redefinition is the bias value at which G varies fastest.

In the third panel of Fig. 8.15, G is plotted for the same case considered in the first panel of Fig. 8.12, which, as we have remarked before, shows

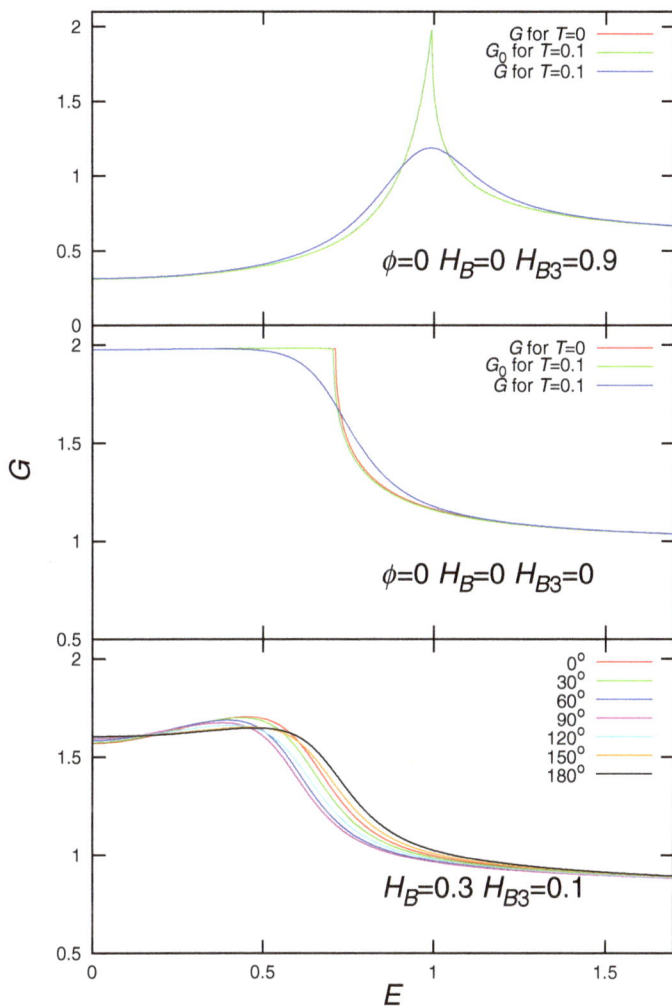

Fig. 8.15 Illustration of contributions to the temperature dependence of $G(E)$. In the top two panels G is plotted at fixed ϕ. The thicknesses and fields are as in Fig. 8.10. Results at $T = 0.1$, in units of T_c^0, are compared to $T = 0$ results. The result of including only G_0, the $T = 0, 1$ correction to G arising from the T dependence of $\Delta(y)$ is also shown, and it is seen to be nearly identical to the zero T value, particularly in the top panel. The top panel is for a very high barrier ($H_{B3} = 0.9$) between S and F_2 and $H_{B1} = H_{B2} = 0$, while in the second all $H_{Bi} = 0$. The last panel illustrates (for the same values as the first panel in Fig. 8.12), a case where the CB varies very non-monotonically with angle, and shows how little this behavior is affected by T.

good spin valve effects in its CB properties, but now at $t = 0.1$ instead of at zero temperature. The two results should be carefully compared. While the curves are now much smoother the behavior of the different features with angle are robust. In particular the sharp minimum of the critical bias at $\phi = 90°$ remains unchanged. This [30] is generally the case. Hence, spin valve properties are only weakly dependent on T.

8.5 Summary

This has been a long chapter — the longest in the book. In it, we have seen how to calculate in a practical and efficient way, the conductance of many SF structures via a transfer matrix method. As illustration, we have seen in many figures a variety of results for $G(E)$ with emphasis on the spin valve effects that occur.

Many of the results show an oscillatory and even apparently resonant structure as certain geometrical and physical parameters are varied. The physical reasons for this are at this point not clear. We will see in Chap. 10 that many are related to the interplay between spin and charge transport. But to see this, we must first learn about spin transport, which is of course a very important subject in its own right. This we will do next, in Chap. 9.

Chapter 9

Spin Transport

9.1 Introductory Remarks

In the previous two chapters we have been looking at charge transport in ferromagnet/superconductor heterostructures. First, in Chap. 7 we considered the general methodology and basic ideas, and then in Chap. 8 we studied in considerable detail the bias dependence of the conductance in FFS and $FNFS$ spin valve structures, as obtained via a transfer matrix method. But this is obviously not enough: as the electrons and holes move around and are transmitted and reflected by several processes, some ordinary, some peculiar, they carry with them not only their charge, but also their spin. Furthermore, although ordinary singlet Cooper pairs do not carry net spin, triplet pairs with $m = \pm 1$ do and we have seen, in Chap. 5, how they are generated. We have also seen, in Sec. 6.3.2, that their existence is quantitatively verified experimentally. Thus, spin is transported and the mechanisms for its transport are physically nontrivial. Also, we have already indicated in Sec. 8.4.2 that part of the conductance vs. bias peak structure is due to spin transport.

The above implies three things: the first is that spin transport in these structures will yield very valuable information about the physical processes present. The second, and less obvious at this point, is that from the properties of spin transport inferences can be made regarding possible applications to spintronics. The third is that consideration of spin transport will help us better understand aspects of charge transport.

9.2 Basic Ideas

The first basic thing that we need is the expression for the spin current and its basic properties.

Now, spin is a vector. Therefore spin current is a tensor. This is just as in elementary physics: momentum current is a tensor of rank two- the stress tensor in hydrodynamics, Maxwell's stress tensor in electromagnetism. It is not really difficult to write the full expression for this spin current tensor in any geometry. However, in the quasi one dimensional situation things are simpler. This we can see from considering the charge current: it is of course a vector, but in the quasi one dimensional case it has only one component, j_y, which furthermore depends only on y, the direction normal to the layers. Having only one component means in effect that it can be dealt with as a scalar, and its expression is then given by Eq. (7.8). Consider now the spin current. In its rank two tensorial nature one of the two indices corresponds to spin and one to space: one is looking, e.g., to the x component of the spin being transported in the spatial y direction. When considering the simplifications that occur in the layered geometry, we see that the spin index must remain: with all the varying directions of the internal fields, the spin can point in any direction. The other index is the spatial one, and there things do simplify. The current, for each of the three spin components, must have only one spatial component (the one normal to the layers, which is the y component in our notation), and depend only on y. Hence, from this point of view it is a spatial scalar. Therefore the spin current tensor can be characterized, in the quasi one dimensional case, as a *vector* $S_i(y)$ where the index i characterizes the direction of the spin.

Charge and spin currents have different conservation laws. We have seen in Sec. 7.2 that the charge current density must be a constant throughout the system, in the steady state. In contrast, we will see below that the spin current density is generally not a conserved quantity in the ferromagnetic regions, because of the existence of torques.

To derive expressions for the spin current, an important quantity to start with is the local magnetization **m** given by

$$\mathbf{m}(\mathbf{r}) \equiv -\mu_B \sum_\sigma \langle \psi_\sigma^\dagger \boldsymbol{\sigma} \psi_\sigma \rangle, \tag{9.1}$$

where μ_B is the Bohr magneton. We have briefly discussed this quantity in Sec. 4.3.4 in connection with an example of the reverse proximity effect in SFS trilayers, but here we need a more general expression. Rewriting the above equation a bit more carefully, taking into account the finite bias as in Eqs. 7.7 or 7.8 we have the non-equilibrium local magnetization components:

$$m_i \equiv \sum_{\epsilon_{\mathbf{k}}<eV} \sum_{\sigma} -\mu_B \langle \psi_\sigma^\dagger \sigma_i \psi_\sigma \rangle_{\mathbf{k}}. \tag{9.2}$$

By performing a Bogoliubov transformation, Eq. (2.5), in Eq. (9.2) one gets, with a bit of care, in the low-T limit, for each of the three components:

$$m_x = -\mu_B \left[\sum_n \left(-v_{n\uparrow} v_{n\downarrow}^* - v_{n\downarrow} v_{n\uparrow}^* \right) \right.$$
$$\left. + \sum_{\epsilon_{\mathbf{k}}<eV} \left(u_{\mathbf{k}\uparrow}^* u_{\mathbf{k}\downarrow} + v_{\mathbf{k}\uparrow} v_{\mathbf{k}\downarrow}^* + u_{\mathbf{k}\downarrow}^* u_{\mathbf{k}\uparrow} + v_{\mathbf{k}\downarrow} v_{\mathbf{k}\uparrow}^* \right) \right] \tag{9.3a}$$

$$m_y = -\mu_B \left[i \sum_n \left(v_{n\uparrow} v_{n\downarrow}^* - v_{n\downarrow} v_{n\uparrow}^* \right) \right.$$
$$\left. - i \sum_{\epsilon_{\mathbf{k}}<eV} \left(u_{\mathbf{k}\uparrow}^* u_{\mathbf{k}\downarrow} + v_{\mathbf{k}\uparrow} v_{\mathbf{k}\downarrow}^* - u_{\mathbf{k}\downarrow}^* u_{\mathbf{k}\uparrow} - v_{\mathbf{k}\downarrow} v_{\mathbf{k}\uparrow}^* \right) \right] \tag{9.3b}$$

$$m_z = -\mu_B \left[\sum_n \left(|v_{n\uparrow}|^2 - |v_{n\downarrow}|^2 \right) \right.$$
$$\left. + \sum_{\epsilon_{\mathbf{k}}<eV} \left(|u_{\mathbf{k}\uparrow}|^2 - |v_{\mathbf{k}\uparrow}|^2 - |u_{\mathbf{k}\downarrow}|^2 + |v_{\mathbf{k}\downarrow}|^2 \right) \right], \tag{9.3c}$$

where the first summations in the expressions for m_i denote the ground state local magnetizations. The second summations appear as a consequence of the finite bias: the bias changes the quasiparticle and quasihole motions, and the spin is dragged along with the charge. Notice that all three components can be nonzero, even in this quasi one dimensional limit: the local magnetization is always a vector in spin space.

We are now ready to write the continuity equation for the local magnetization just as we did for the charge current, see Sec. 7.2. By using the Heisenberg equation

$$\frac{\partial}{\partial t} \langle \mathbf{m}(\mathbf{r}) \rangle = i \langle [\mathcal{H}_{eff}, \mathbf{m}(\mathbf{r})] \rangle \tag{9.4}$$

one obtains the relation:

$$\frac{\partial}{\partial t}\langle m_i \rangle + \frac{\partial}{\partial y} S_i = \tau_i, \quad i = x, y, z \tag{9.5}$$

where τ is a *torque*, called the spin-transfer torque (STT):

$$\boldsymbol{\tau} \equiv 2\mathbf{m} \times \mathbf{h}, \tag{9.6}$$

and the the components of the spin current (strictly speaking the spin current density) S_i are given by

$$S_i \equiv \frac{i\mu_B}{2m} \sum_\sigma \left\langle \psi_\sigma^\dagger \sigma_i \frac{\partial \psi_\sigma}{\partial y} - \frac{\partial \psi_\sigma^\dagger}{\partial y} \sigma_i \psi_\sigma \right\rangle. \tag{9.7}$$

From Eq. (9.5), we can see that $\mathbf{S}$ is a local physical quantity and the STT, τ, is responsible for the change of local magnetizations due to the flow of spin-polarized currents. The conservation law Eq. (9.5) (with the spin torque term) for the spin density is as fundamental as the conservation law for the charge and one has to check it is not violated when studying these transport quantities.

The expressions for the nonequilibrium spin current components can again be rewritten to carefully include the bias dependence as:

$$S_i \equiv \frac{i\mu_B}{2m} \sum_{\epsilon_\mathbf{k} < eV} \sum_\sigma \left\langle \psi_\sigma^\dagger \sigma_i \frac{\partial \psi_\sigma}{\partial y} - \frac{\partial \psi_\sigma^\dagger}{\partial y} \sigma_i \psi_\sigma \right\rangle_\mathbf{k}, \tag{9.8}$$

which, upon performing the Bogoliubov transformation become, in the low T limit, spelling out each component:

$$S_x = \frac{-\mu_B}{m} \text{Im} \left[\sum_n \left(-v_{n\uparrow} \frac{\partial v_{n\downarrow}^*}{\partial y} - v_{n\downarrow} \frac{\partial v_{n\uparrow}^*}{\partial y} \right) \right.$$

$$\left. + \sum_{\epsilon_\mathbf{k} < eV} \left(u_{\mathbf{k}\uparrow}^* \frac{\partial u_{\mathbf{k}\downarrow}}{\partial y} + v_{\mathbf{k}\uparrow} \frac{\partial v_{\mathbf{k}\downarrow}^*}{\partial y} + u_{\mathbf{k}\downarrow}^* \frac{\partial u_{\mathbf{k}\uparrow}}{\partial y} + v_{\mathbf{k}\downarrow} \frac{\partial v_{\mathbf{k}\uparrow}^*}{\partial y} \right) \right] \tag{9.9a}$$

$$S_y = \frac{\mu_B}{m} \text{Re} \left[\sum_n \left(-v_{n\uparrow} \frac{\partial v_{n\downarrow}^*}{\partial y} + v_{n\downarrow} \frac{\partial v_{n\uparrow}^*}{\partial y} \right) \right.$$

$$\left. + \sum_{\epsilon_\mathbf{k} < eV} \left(u_{\mathbf{k}\uparrow}^* \frac{\partial u_{\mathbf{k}\downarrow}}{\partial y} + v_{\mathbf{k}\uparrow} \frac{\partial v_{\mathbf{k}\downarrow}^*}{\partial y} - u_{\mathbf{k}\downarrow}^* \frac{\partial u_{\mathbf{k}\uparrow}}{\partial y} - v_{\mathbf{k}\downarrow} \frac{\partial v_{\mathbf{k}\uparrow}^*}{\partial y} \right) \right] \tag{9.9b}$$

$$S_z = \frac{-\mu_B}{m} \mathrm{Im} \left[\sum_n \left(v_{n\uparrow} \frac{\partial v_{n\uparrow}^*}{\partial y} - v_{n\downarrow} \frac{\partial v_{n\downarrow}^*}{\partial y} \right) \right.$$

$$\left. + \sum_{\epsilon_\mathbf{k} < eV} \left(u_{\mathbf{k}\uparrow}^* \frac{\partial u_{\mathbf{k}\uparrow}}{\partial y} - v_{\mathbf{k}\uparrow} \frac{\partial v_{\mathbf{k}\uparrow}^*}{\partial y} - u_{\mathbf{k}\downarrow}^* \frac{\partial u_{\mathbf{k}\downarrow}}{\partial y} + v_{\mathbf{k}\downarrow} \frac{\partial v_{\mathbf{k}\downarrow}^*}{\partial y} \right) \right].$$

(9.9c)

Once more, the first summations on the right side of Eq. (9.9) represent the static spin current densities when there is no bias. The static spin current does not need to vanish, since a static spin-transfer torque may exist near the boundary of two magnets with misaligned exchange fields. On the other hand, the spin-transfer torque has to vanish in the superconductor or the normal layers, where the exchange field is zero. The finite bias, the second sums on the right side of Eq. (9.9) leads to a non-equilibrium quasi-particle distribution for the system and results in non-static spin current densities.

It is conventional to normalize $\mathbf{m}$ (see Sec. 4.3.4) via $-\mu_B(N_\uparrow + N_\downarrow)$, where the number densities $N_\uparrow = k_{FS}^3(1 + h_m)^{3/2}/(6\pi^2)$ and $N_\downarrow = k_{FS}^3(1 - h_m)^{3/2}/(6\pi^2)$. Following this convention, one normalizes $\boldsymbol{\tau}$ to $-\mu_B(N_\uparrow + N_\downarrow)E_{FS}$ and $\mathbf{S}$ to $-\mu_B(N_\uparrow + N_\downarrow)E_{FS}/k_{FS}$.

9.3 Spin Current Conservation Laws

As we have discussed in Chap. 8 in connection with charge transport, great care has to be taken, in any transport calculation, to make absolutely sure that the proper conservation laws are satisfied. In spin transport calculations the "holy" rule that cannot be broken is the relation Eq. (9.5). Actually, since we are here in a steady state, that equation takes the form:

$$\frac{\partial}{\partial y} S_i = \tau_i, \quad i = x, y, z,$$

(9.10)

which must be satisfied in all the computations.

One may naively think that the right hand side of Eq. (9.10) should be zero: after all it is given by Eq. (9.6) and one might expect to have the vectors $\mathbf{m}$ and $\mathbf{h}$ to be parallel to each other. This would be a sensible conclusion if all the internal fields were aligned. But if there are two F layers and the misalignment angle ϕ between the two fields $\mathbf{h_1}$ and $\mathbf{h_2}$ was not zero, and in general it is not, then the magnetization near the interface between the two ferromagnets, or between one of them and an N layer, can be expected to be effectively rotated. It will be verified below that this is indeed the case. So there will be a torque, called the spin-transfer torque (STT), which will be particularly strong near the interfaces and will affect the spin currents.

If results obtained showed a violation of Eq. (9.10) they would be incorrect. Where would the mistake come from? Well, lack of self-consistency might be, as in the charge case, a problem. But it is also possible to have numerical complications of other kinds. This is because for spin transport, because of the STT, the full spin currents are a function of position. Hence one is not really looking for a characterization in terms of a transport coefficient such as some kind of spin conductance tensor. Rather, one needs to calculate the position dependent spin currents themselves. Furthermore there is no intuitive shortcut here such as the one that led us to Eq. (8.21) for example. One has no real choice but to calculate the spin current from the full equations Eq. (9.9). For charge transport, one could also alternatively calculate the current from the corresponding expression, Eq. (7.8), and extract the conductance by differentiation with respect to V. But it is much more complicated. And certainly one would never bother to study the current as function of position, since the plot would be completely boring and flat. For the spin current the spatial dependence is nontrivial, as we shall see. Therefore, for spin, one must perform a much more complicated calculation, which involves doing the sums in Eq. (9.9) numerically. Mistakes can occur and one should always prudently check that Eq. (9.10) is satisfied.

To clarify how to proceed now, recall the method used in Chap. 8 to calculate the conductance $G(V)$. It involved merely evaluating the reflection and transmission amplitudes governed by the continuity of the wavefunction and discontinuity of its derivatives (for both particle and hole, and for each spin, i.e., including both ordinary and Andreev reflections) at each interface, as one would do in elementary quantum mechanics. In the S electrode, the procedure involved dividing it into arbitrarily thin layers, in each of which the y-dependent self-consistent pair potential, as determined numerically, could be replaced by a constant. As mentioned in Chap. 8, specifically in the discussion in the second paragraph following Eq. (8.20), the transfer matrix method yields, in addition to the transmission and reflection amplitudes used to evaluate G, the self-consistent wavefunctions that satisfy the proper transport boundary conditions. For charge transport, we used those only as a consistency check. Here things are different.

In the expressions of the local magnetization, Eq. (9.3), and the spin current, Eq. (9.9), we have, as we have said, two terms in the right sides.

The first is the equilibrium result, which can be calculated straightforwardly by the methods of Chap. 2. The more important terms are, of course, the bias-driven contributions. To evaluate those one uses the "rebuild" wave-functions as explained in the previous paragraph. These wavefunctions correspond to the proper boundary conditions of injected spin up or spin down particles (see e.g., Eq. (8.1) and Eq. (8.2) in the Sec. 8.2 discussion). The method is, in its bare conceptual essence, nothing but the elementary quantum mechanical procedure of building plane wave solutions out of stationary state wavefunctions, but here it is mathematically much more complicated. The transfer matrix method simply transcribes the continuity conditions for the wavefunctions, and the discontinuity in the derivatives arising from the delta function interfacial scattering, to each adjacent layer. From these rebuilt wavefunctions the second terms in the right sides of the expressions for $\mathbf{m}(y)$ and $\mathbf{S}(y)$ are straightforwardly calculated by adding the appropriate terms. This procedure is especially important in spin transport calculations, as the quantities involved depend on position and the simple BTK [25] procedure that one employs for the conductance does not apply.

9.4 Spin Transport Results

The results of this section can be summarized as follows: The finite bias leads to spin currents. As opposed to the ordinary charge currents, these spin currents are generally not conserved locally because of the presence of the spin-transfer torques which act as source terms and are responsible for the change of spin-density. But a self-consistent calculation *must* still contain exactly the correct amount of non-conservation, that is, Eq.(9.10) must be satisfied. It is therefore of fundamental importance to verify that it is, as we have. The plots shown in this chapter are taken from Refs. [29, 30, 32]. The currents are studied as a function of position, as well as of bias, geometry, and material parameters.

9.4.1 *The FFS trilayer*

As in Chap. 8 I will begin with some results for FFS trilayers although the emphasis in this chapter will be heavily on the experimentally relevant $FNFS$ spin valve structure.

Let us first consider, in the next two figures, spin transport in a trilayer with $h = 0.1$ and a non-collinear orthogonal magnetic configuration, $\phi = 90°$. Thus, the internal field in the outer electrode F_1 is along the z-axis, while that in F_2 is along x, (see Fig. 2.1). The thicknesses are $D_{F1} = 250$, $D_{F2} = 30$, and $D_S = 250 = 5\Xi_0$. For these two figures the interfaces are ideal.

A set of results is shown in Fig. 9.1. There, in the three main plots, are shown the three components of the spin current density, computed from Eq. (9.9) and normalized as explained. They are plotted as functions of the dimensionless position $Y \equiv k_F y$ for several values of the bias $E \equiv eV/\Delta_0$. The F_2/S and the F_1/F_2 interfaces are located at $Y = 0$ and $Y = -30$, respectively. For clarity, only the range of Y corresponding to the "central" region near the interfaces is included in these plots: the shape of the curves deeper into S or F_1 can be easily inferred by extrapolation.

From these main panels, one sees that beyond (to the right) of the F_1/F_2 interface, that is, not only in the F_2 but also in the S region, the spin current polarization is in the x-direction, the direction of the exchange field in F_2. Furthermore, S_x is a constant in space except in the F_1 region, where it exhibits oscillatory behavior. This indicates the existence of a non-vanishing, oscillating spin-transfer torque in the F_1 layer, as we will verify below. Also, S_x vanishes everywhere when the bias is less than the super-conducting gap in bulk S ($E < 1$). Thereafter, its magnitude, including its constant value in F_2 and S, increases with bias. In fact, the behavior of S_x with E is reminiscent of that of the ordinary charge current in an N/S junction with a very strong barrier (tunneling limit) where there is no current until $eV > \Delta_0$, that is, $E > 1$.

The S_y component, along the normal to the layers, is shown in the middle main panel of Fig. 9.1. It depends extremely weakly on the bias E. It is very small except near the interface between the two ferromagnets but there it is about an order of magnitude larger than the other two components. Because of this only a somewhat smaller Y range is included in the plot. Unlike the S_x and S_z components, S_y does not vanish even when there is no bias applied to the trilayer. From this observation, one can infer that the value of S_y is largely derived from its static part with only a very small contribution from the effect of finite bias. The emergence of a static spin current is due to the leakage of the local magnetization m_z into the F_2 layer and of m_x into the F_1 layer, leading to a spatially sharp y component of the torque. This component, which will be seen below in Fig. 9.2, leads via Eq. (9.5) to a static S_y largely localized near the F_1/F_2 interface. The S_z component (lower panel) is constant in the F_1 region, as one would expect, with a bias dependence there similar to that S_x has in the F_2 layer, except that, as opposed to the S_x component, it is non-vanishing, although very small, when $E < 1$. It increases rapidly with bias when

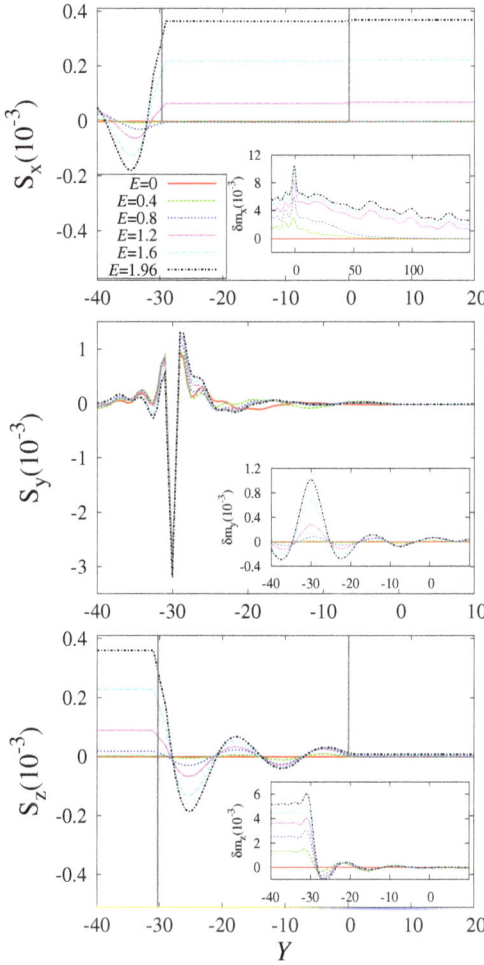

Fig. 9.1 Spin current density components, S_x, S_y, and S_z, Eq. (9.9), vs. Y for several values of E (main panels). In the plots, $\phi = 90°$, $h = 0.1$, $D_{F1} = 250$, $D_{F2} = 30$, $D_S = 250 = 5\Xi_0$. The F_2/S interface is at $Y = 0$ and the F_1/F_2 interface at $Y = -30$. Vertical lines help locate the layers. Only a portion of the Y range is included (see text). Insets: change in the local magnetization, $\delta\mathbf{m}(E) \equiv \mathbf{m}(E) - \mathbf{m}(0)$. Values of E as in the main plot, the Y ranges may be different. The magnitude of the x and z components of both $\delta\mathbf{m}$ and $\mathbf{S}$ increases with E,

$E > 1$. S_z oscillates in the F_2 region, due to the torque there, as we will verify below. It vanishes in the S layer.

In summary, the behavior of the spin current vector, in this $\phi = 90°$ configuration, is as follows: when $E > 1$, the spin current, which is initially

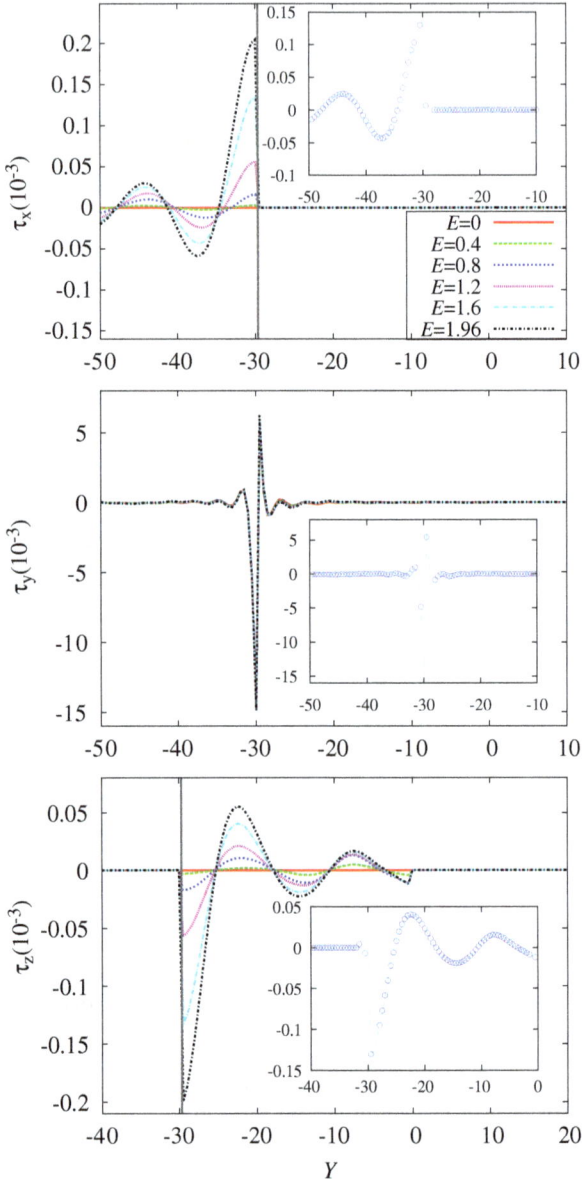

Fig. 9.2 Components of $\boldsymbol{\tau} \equiv 2\mathbf{m} \times \mathbf{h}$ vs. Y for several values of E, indicated. Parameters and geometry are as in Fig. 9.1. Insets: (for $E = 1.6$) torque (dashed line) and derivative of spin current density (circles). The calculations uphold Eq. (9.10)).

(at the left side) spin-polarized in the $+z$ direction (see Fig. 2.2), is twisted to the x direction under the action of the spin torques as it passes through the second magnet. The STT is discussed in more detail below, This second magnet, therefore, acts as a *spin filter*. The current remains then with its spin polarization in the $+x$ direction as it flows through the superconductor. Thus in this range of E the trilayer *switches the polarization* of the spin current. On the other hand, when $E < 1$, the small z-direction spin-polarized current tunneling into the superconductor is gradually converted into supercurrent and becomes spin-unpolarized.

In the insets of the three panels of Fig. 9.1, one can see the behavior with bias of the corresponding component of the local magnetization as it is carried into S, dragged, so to speak, by the spin and charge currents. Specifically, the components of the vector difference between the local magnetization with and without bias $\delta\mathbf{m}$ are plotted as a function of Y. This local quantity is specifically defined as

$$\delta\mathbf{m}(E) \equiv \mathbf{m}(E) - \mathbf{m}(E = 0), \qquad (9.11)$$

and can be viewed as a measure of the local spin accumulation: it obviously depends on Y. The range of Y included is again chosen to display the salient aspects of the behavior of this quantity, and it is not the same as in the main plots, nor is it the same for each component. The bias values are the same as in the main plots, however, and so is everything else. The magnetizations are computed from Eq. (9.3) and normalized in the usual way as explained at the end of Sec. 9.2. In these units, at $h = 0.1$ the value of the dominant component of $\mathbf{m}$ in the magnetic layers is about 0.15. This scale should be kept in mind.

The behavior of the x component in the F_2 and S regions is nontrivial, and the corresponding Y range is what is included in the top panel inset. When the applied bias is below the bulk S gap value, $\delta m_x(E)$ penetrates into the S layer with a decay length $\sim \Xi_0 = 50$. This decay length is much longer than that found for the static magnetization, $\mathbf{m}(0)$ in Chap. 4. When the bias is larger, $E > 1$, $\delta m_x(E)$ penetrates even more deeply into the S layer, with a behavior clearly very different than that for $E < 1$. This long-range propagation is of course consistent with the behavior of S_x, as S_x, the spin current component in the x direction, appears only when $E > 1$. The magnitude of δm_y (inset in second panel) is much smaller than that of δm_x or δm_z. It peaks near the F_1/F_2 interface and that range of Y is emphasized in the middle inset. Its overall scale monotonically increases with increasing bias. It dampens away from the F_1/F_2 interface in an oscillatory manner. The component δm_z, which can conveniently be plotted in the same Y range as done in the bottom panel inset, decays with a very short decay length and oscillates in F_2. The overall damped oscillatory behavior of δm_y and δm_z in the F_2 region reflects the precession, as a function of position, of the spin density around the local exchange field, which points along the $+x$ direction. This phenomenon is well known in spin valves. The spatial oscillation periods for δm_y, δm_z, S_x, and S_z are very similar and of the order of $1/h$, as could be expected from the findings in Sec. 2.7.

Next, let us look at the spin-transfer torque, Eq. (9.6). The components of this vector quantity, computed from the normalized values of h and $\mathbf{m}$, are plotted as a function of position in Fig. 9.2 for exactly the same parameters as in Fig. 9.1. Note that the chosen normalizations of the dimensionless $\mathbf{h}$ and $\mathbf{m}$ determine how $\boldsymbol{\tau}$ is normalized. Results are shown for each of the three components of $\boldsymbol{\tau}$ in the main panels of the figure. Again, only relevant portions of the Y range are included and vertical lines indicate interface locations at $Y = -30$ and $Y = 0$. One sees that at zero bias, $E = 0$, both τ_x and τ_z vanish identically. In the F_1 layer, τ_x increases in magnitude with increasing E. It vanishes in F_2 and in S. The behavior of τ_z is, as one would expect, the converse: it vanishes in F_1 (and in S), while its magnitude increases in F_2. The oscillatory behavior of τ_x and τ_z is completely consistent, as we shall see below, with the results for S_x and S_y. The component normal to the layers, τ_y, is nonvanishing only near the F_1/F_2 interface, although its peak there attains a rather high value, nearly two orders of magnitude larger than the peak value of the other components. It is nearly independent of bias, which is in agreement with our previous discussion on the behavior of S_y.

In the insets, it is verified for each component, that Eq. (9.10) is satisfied, that is, that the self-consistent methods used in this book strictly preserve the conservation laws in this very nontrivial case. Specifically the bias value $E = 1.6$ is considered as an illustration. Examine first the top panel inset. There are plotted both the x component of the spin-transfer torque, τ_x (dashes), taken from the corresponding main plot, and the spatial derivative of the spin current, $\partial S_x/\partial Y$ (circles), obtained by numerically differentiating the corresponding result in the top panel of Fig. 9.1. Clearly, the curves are in perfect agreement. One can easily check that with the normalizations and units chosen there should be no numerical factor between the two quantities. In the second panel, the same procedure is performed for the y component, although in this case, because of the very weak dependence of both S_y and τ_y on bias, the value of the latter is hardly relevant. Nevertheless, despite the evident difficulty in computing the numerical derivative of the very sharply peaked S_y, the agreement is excellent. For τ_z, its vanishing in the F_1 region is in agreement with the constant spin current in that layer. The conservation law Eq. (9.10) is verified in the inset for this component, again at bias $E = 1.6$. Just as for the x component, the dots and the line are on top of each other. Thus the conservation law for each component is shown to be perfectly preserved in the calculations. Such numerical checks should be performed whenever possible.

The previous two figures are limited to $\phi = \pi/2$ and, in the insets, to one bias value. Let us now briefly turn to broader ϕ and additional E dependence, still with $h = 0.1$. The next three plots are for a superconductor thickness of five times the coherence length ($D_S = 250 = 5\Xi_0$) so that the saturated value of $\Delta(y)$ is essentially the same as the bulk S value Δ_0. The F_1 layer ($D_{F1} = 250$) is rather thick while $D_{F_2} = 30$.

Let us examine first the plots in Figs. 9.3 and 9.4. These two figures again show results for the three components of the spin current and of the STT respectively, each under the same conditions (see captions). These quantities are now shown for three values of the bias, E, ranging from below to well above $E = 1$: for each component there is a panel corresponding to each value of E. The curves correspond to different values of ϕ as indicated in the legend. At $\phi = 0$ and $\phi = 180°$ the same conservation laws that preclude singlet to triplet pair conversion imply that the torques vanish. It is evident that there is no point in including the regions of the sample deep inside S or even well inside F_1, so the region plotted is that which includes both interfaces: the S/F_2 interface is located at the origin and that between ferromagnets at $Y = -30$.

The results for y-components are the easiest ones to understand. The behaviors of S_y in the middle panels of Fig. 9.3 and that of τ_y in the corresponding panels of Fig. 9.4 are of course strongly correlated, these quantities being connected as they are via Eq. (9.5). We see that S_y has a very strong peak, in absolute value, at the F_1/F_2 interface. This peak indicates a strong oscillation in τ_y which indeed we see has very sharp peaks, with opposite signs, near the F_1/F_2 boundary, where it vanishes. These peaks reflect the existence of a very strong but short-ranged magnetic proximity effect. They are most prominent near $\phi = 90°$. On the other hand, within F_2 or F_1, S_y is very small and oscillatory. These quantities

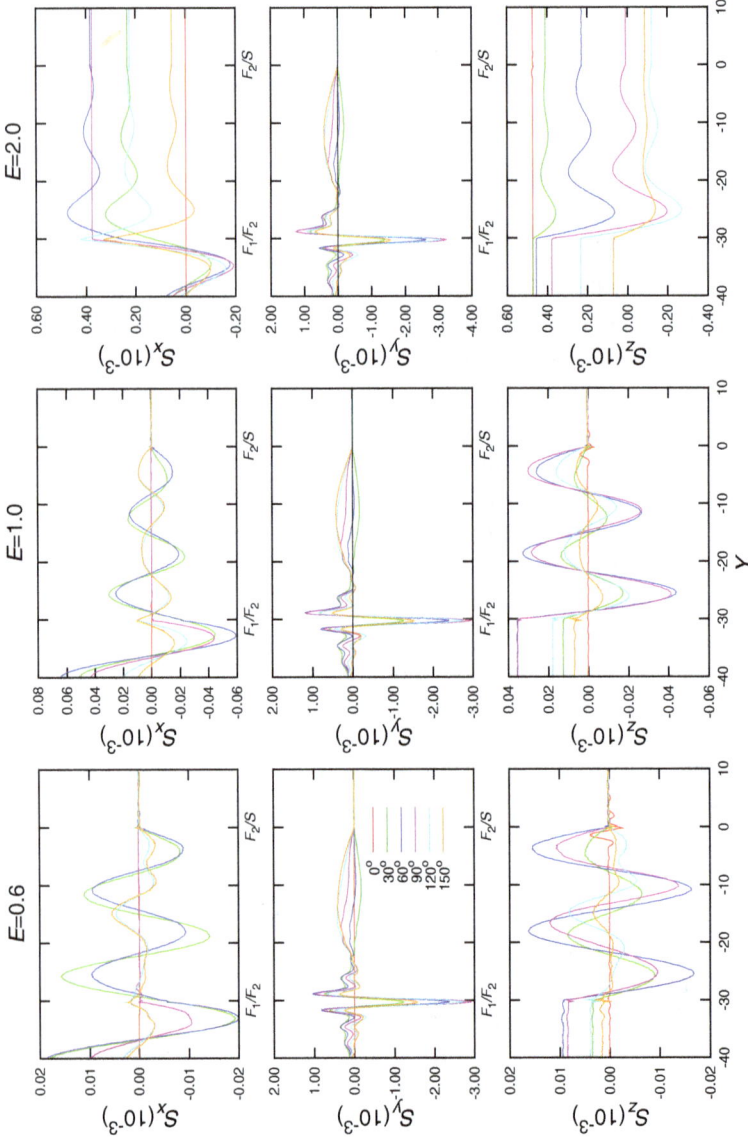

Fig. 9.3 The three components of the spin current shown as a function of Y for several values of ϕ, as indicated, and three values of the bias. We have $h = 0.1$, $D_{F1} = D_S = 250 = 5\Xi_0$, $D_{F2} = 30$. Only the central region of Y is plotted: $Y = 0$ is at the F_2/S interface. All components of $\mathbf{S}$ are zero for $\phi = 0$ (shown) and $\phi = 180°$ (not shown).

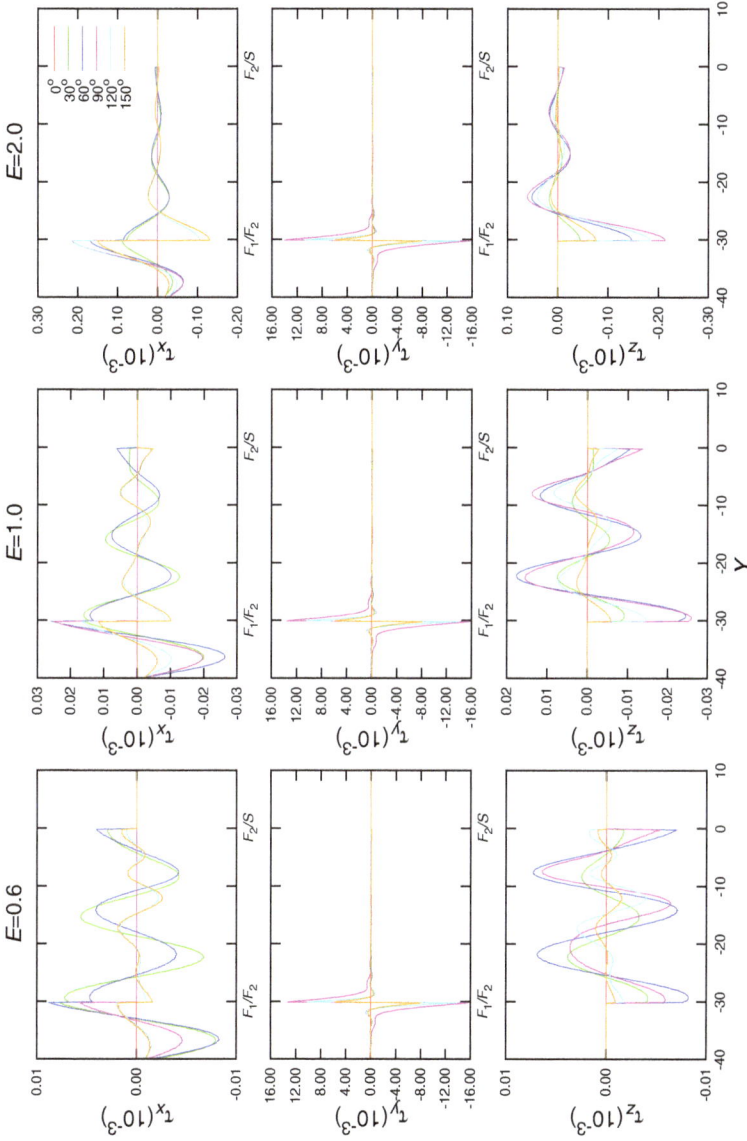

Fig. 9.4 The three components of the spin-transfer torque plotted for the same situation as in Fig. 9.3. The torque is identically zero for $\phi = 0$ and $\phi = 180°$. The discontinuities at the interface reflect those of the internal fields.

depend only weakly on the bias, since they again basically reflect a static effect: the two magnets interacting with each other.

The behaviors of the in-plane components, x and z, are similar to each other (they are after all related by spin rotations) and quite different from that of the y components. Now both currents and torques are transport-induced and one sees immediately that they markedly depend on bias. Since in F_1 the internal field always points along z, we find in Fig. 9.3 that S_z is a constant in F_1, its value increasing with bias. As a function of ϕ its behavior is complicated, the maximum value is near but not at $\phi = 90°$ and its value is dependent on bias. For $\phi = 90°$ the field points along the x direction in F_2 and therefore S_x is independent of Y in F_2 at $\phi = 90°$ only. For other values of the mismatch angle, S_x oscillates in both magnetic layers, and so does S_z in F_2 in every case. The amplitude of the oscillations of S_x decays slowly deep into the F_1 layer. In all cases the period of the spatial oscillations is approximately $1/h$ indicating that the oscillations are due to the behavior of the Cooper pairs. As to the corresponding components of the torque, one notices at once in Fig. 9.4 that their maximum value is much smaller than that of the τ_y peak but, away from the F_1/F_2 interface, the values are not all that different. This reflects the geometry, as explained above. The x and z components of the torque are also nonmonotonic with ϕ, with peaks that are near, but not necessarily at, $\phi = 90°$, depending on the bias. For lower biases, the peak values appear to shift away to smaller values, more closely aligned with the z direction, due to the increasing static effect from the F_1 layer. In the coordinate system used, τ_z vanishes in F_1 for all ϕ and oscillates in F_2. Correspondingly, τ_x is oscillatory in both F_1 and F_2 except at $\phi = 90°$ where it is zero in F_2. The magnetization itself is not displayed in these plots, but its components exhibit damped oscillations which reflect the precession of the magnetization vector around the internal fields. Such precessional behavior then results in the current oscillations discussed above.

In addition to S_z being constant in F_1 and also S_x in F_2 for $\phi = 90°$, all the components of the spin current are trivially constant in the S layer, since there are no torques there. As can be seen in Fig. 9.3, all spin current components vanish in S unless the bias exceeds the bulk S gap, Δ_0. This confirms the remarkable fact mentioned above that, in this one respect, spin currents behave like charge currents in an N/S junction. It can rather easily be shown via standard spin rotation matrix arguments that the constant values of S_z and S_x deep in the S material, in the limit of large bias, should be approximately related to the value of S_z in the F_1 layer by factors

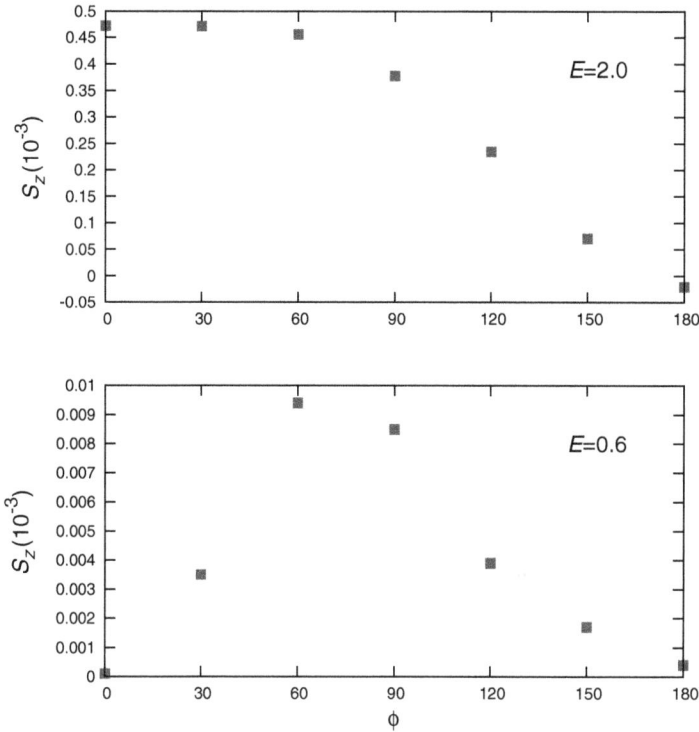

Fig. 9.5 The z component of the spin current in the outer F region as a function of ϕ, at two different bias values. The plots are for the same conditions and parameter values as in Figs. 9.3 and 9.4.

of $\cos\phi$ and $\sin\phi$ respectively, and this can be seen in the last column of Fig. 9.3 to hold rather accurately at $E = 2$. On the other hand, the dependence of the constant value of S_z in the outer layer on ϕ is nontrivial as one can see in Fig. 9.3. This is seen more clearly in Fig. 9.5, where the value of S_z in F_1 at two different bias values is plotted. For values below the CB the behavior is nonmonotonic: it cannot be monotonic, since S_z vanishes at both $\phi = 0$ and $\phi = 180°$. The maximum value is near $\phi = 90°$. On the other hand, when the bias is well above the CB, S_z, which in this case is non-vanishing at zero angular mismatch, decreases monotonically with ϕ. It becomes slightly negative when the two magnets are aligned in opposite directions. The behavior is not described by a simple trigonometric function and a simple argument leading to the behavior seen seems elusive.

9.4.2   *Spin transport in spin valve structures*

We turn now to the spin valve case. The procedure to obtain the results is the same as in the previous subsection except for the presence of the N layer, which can be described as a trivial extension of either the F layer or S layer wavefunction with h or Δ taken to be zero, respectively.

The examples given here are for realistic samples similar to those considered in Chap. 6. It is assumed, therefore, that the two ferromagnetic materials are the same, the field strengths h_1 and $h_2 = h$ are equal. They are taken to be $h = 0.145$ in dimensionless units. This is the value used in Chap. 6 to describe the transition temperature experiments in which Co was the ferromagnetic material. Similarly, the scattering strengths for the two N/F interfaces will be assumed to be the same $H_{B1} = H_{B2} \equiv H_B$. The effective coherence length of the superconducting order parameter is taken to be $\Xi_0 = 115$ which was found to be appropriate for samples in which the S layer was Niobium.

The superconducting layer thickness is set to $D_S = 180$ in all the results, which is large enough compared to Ξ_0 to allow for robust superconductivity, but not so large that the proximity effect would be negligible within the superconductor. It is also assumed that any band mismatch parameters are unity. Although this is not true in real systems, in practice we have seen in Chap. 6 that the effects of such a mismatch can be incorporated into the effective value of the scattering strength parameter when interpreting and fitting data.

In this and the following subsections we will examine carefully the spin transport in these structures as a function of geometry and sample quality. Six different sets of physical parameters for the scattering barriers H_{Bi} and for the intermediate layer thicknesses D_N and D_{F2} will be considered. The thickness of the outer F_1 will be kept constant and relatively large. For each set of parameters we will examine the following vector quantities: the spin current, the spin-transfer torque (STT), and the magnetization. For the last one we will focus on the bias dependent local magnetization deviation, as defined in Eq. (9.11), which we denote here the "spin accumulation". Results for the current and for the local spin accumulation as defined in Eq. (9.11) will be examined, as a function of Y, at low-bias $E = 0.6$ and high-bias values $E = 2$. We will focus also on some averaged quantities such as the average spin accumulation in S and N, and the average STT in both F layers. The quantities δm_i and τ_i averaged over one layer will be shown as a function of the bias. Averages over one layer are defined in a rather obvious way, for example:

$$\langle \tau_i \rangle \equiv \frac{1}{D_I} \int dY \tau_i(Y) \tag{9.12}$$

where the integral is over the relevant layer, of thickness D_I. In all cases the dependence of the the results on the angular mismatch angle ϕ of the exchange fields is examined. It would take a book much thicker than this one to examine all relevant quantities for each set of physical parameters, therefore we focus on only the most remarkable features and angular dependencies, and on their distinctive behavior as a function of the physical parameters.

9.4.3 *Spin transport in spin valve structures: Interfacial scattering*

9.4.3.1 *The ideal case*

We consider first, for pedagogical reasons, the case of ideal interfaces, where the proximity effects are maximized. The fundamental features of each quantity studied can be taken as a baseline for comparison with subsequent figures. In Figs. 9.6–9.9 the layer thicknesses for the $F_1 N F_2 S$ layers are $30/40/25/180$ respectively, and both H_B, the barrier strength for both the F/N interfaces, and H_{B3}, the barrier strength for the F_2/S interface are zero. The results can be compared with those shown in the previous subsection for the case where the normal metal layer N is absent. This normal layer of course reduces the STT at the interfaces between the ferromagnets.

To illustrate the usefulness of the spatial averaging, I begin by showing, in Fig. 9.6, plots of the STT, averaged over either of the F layers of an ideal spin valve structure, plotted as a function of bias, E. Only the ferromagnetic regions are included in the plot, since the torque vanishes elsewhere. The component τ_z is zero in the outer ferromagnetic region F_1, since the field h_1 is in the z direction, and the plot of τ_z for that region is therefore not included. The torque τ is always zero for $\phi = 0$ and $\phi = 180°$, and $\tau_x = 0$ for $\phi = 90°$. This is also true for all cases we analyze later in this section. There is clearly a strong critical bias feature in the x and z components in both ferromagnetic regions: the torque is zero below CB, and then grows linearly with increasing bias. The x component in F_1, and the z component in F_2, show similar behavior, with a steady increase or decrease in value respectively for all angles, and a maximum magnitude between

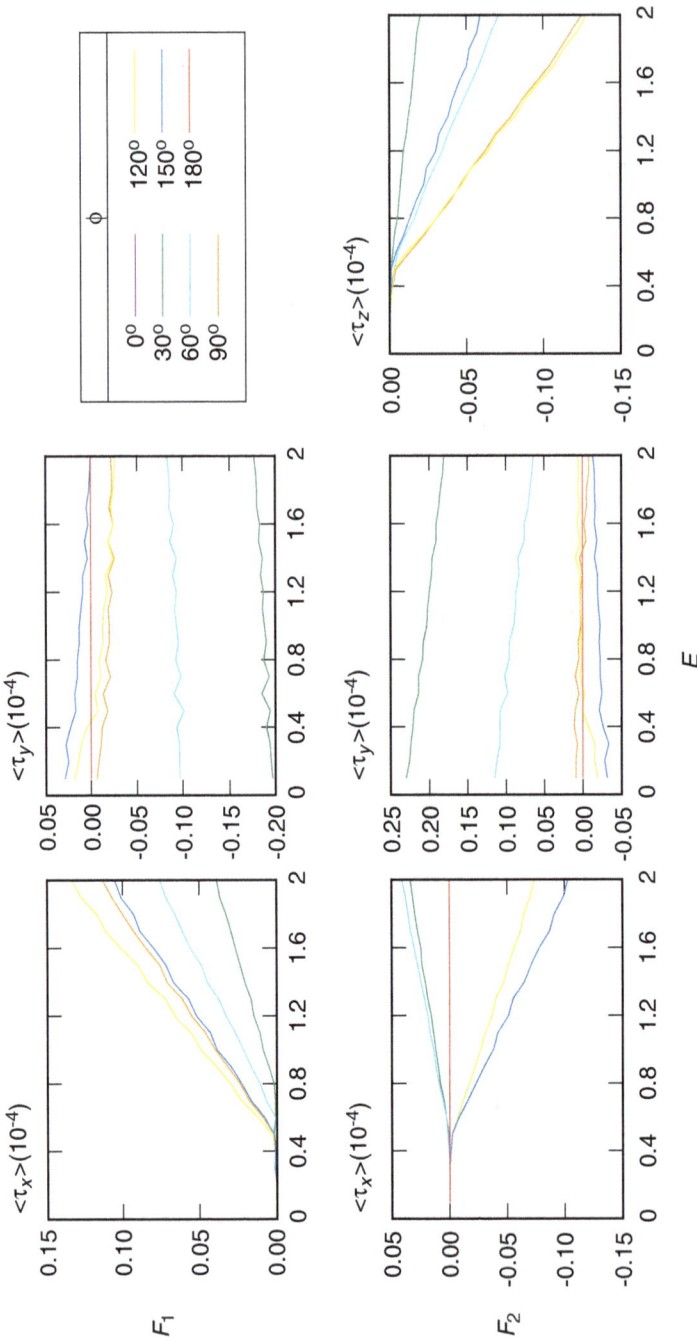

Fig. 9.6 The three components of the STT averaged (see Eq. (9.12)) over the two F layers of a spin valve $F_1 N F_2 S$ structure with ideal interfaces. They are plotted as a function of bias E. Results are given for several values of the misalignment angle ϕ as given in the key in the upper right. The torque is identically zero for $\phi = 0$ and $\phi = 180°$. The layer thicknesses are 30/40/25/180 respectively.

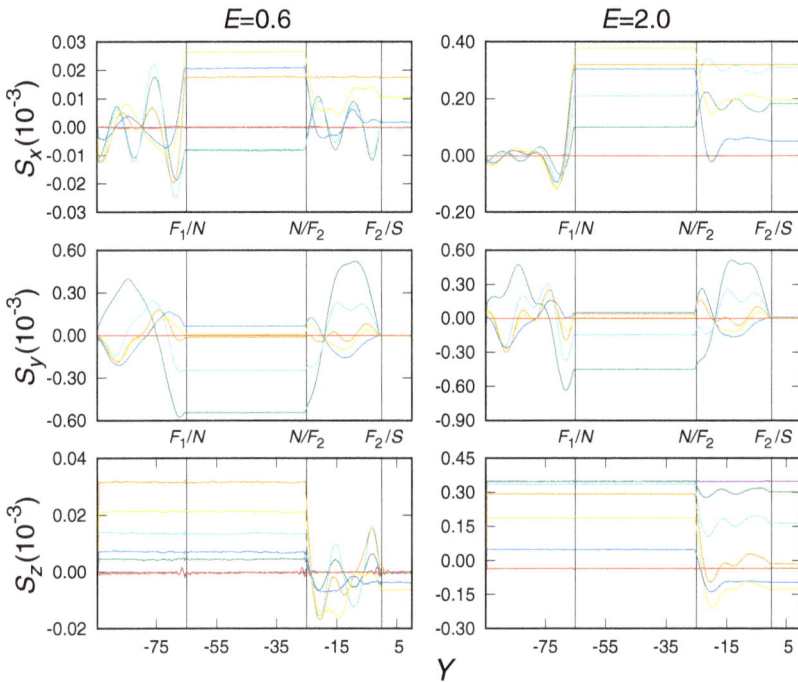

Fig. 9.7 Spin current for the same conditions and color key as in Fig. 9.6. The spin current components are plotted versus Y for several values of the misalignment angle ϕ, as indicated in the third panel of Fig. 9.6. The location of the interfaces is indicated. Most of the range within the S layer is not included in the plot. See text for details.

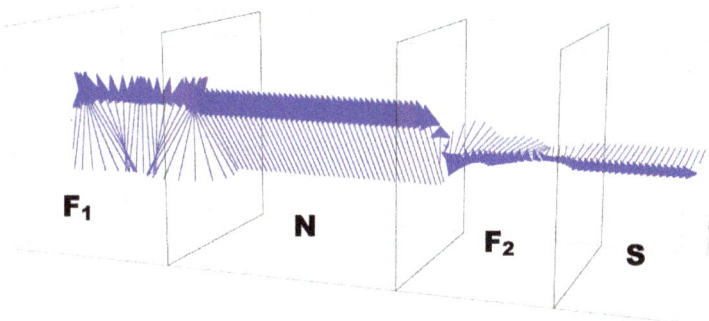

Fig. 9.8 A 3D representation of the spin current from Fig. 9.7 at $\phi = 90°$ and $E = 2.0$. From left to right, the figure comprises the layers $F_1 N F_2 S$ respectively. The spin current precesses about the exchange field in the F layers, with a dampening amplitude in F_2. The orientation of the current in S is rotated to $90°$.

$\phi = 90°$ and $\phi = 120°$. However, τ_x in F_2 is different: it becomes more symmetrical with increasing E for angles $\phi < 90°$ and decreasing for angles $\phi > 90°$. The longitudinal component τ_y has a very different behavior, and is nonzero at zero bias due to the static ferromagnetic reverse proximity effect. Because of this, τ_y is (s in the FFS case) nearly independent of bias, slightly decreasing in magnitude in both ferromagnetic regions. It follows from Eq. 9.5 in the steady state that the net change in spin current in N and S is directly proportional to the average torque. Indeed, the constant S_y in the normal metal can be described by the net result of the average torque τ_y in both ferromagnetic regions.

Next, in Fig. 9.7, we see the spin current components S_i (top to bottom) as a function of position at one lower value of the bias (left) and one above the CB (right). Only the relevant parts of the Y range are included and the position of the interfaces is indicated. In each panel one can see the dependence of the spin current on ϕ (see Fig. 9.6 for the key). In each case the spin current components at $\phi = 0$ and $\phi = 180°$ are constant, as there are no spin torques when $\mathbf{h}_1$ and $\mathbf{h}_2$ are collinear. Furthermore, S_x for $\phi = 90°$ is constant in F_2 since $\mathbf{h}_2$ in this case is along the x-axis. Similarly, S_z is constant for all ϕ in F_1 since $\mathbf{h}_1$ is along the z-axis. At larger bias values, the magnitude of the spin current increases, except for the y-component, normal to the layers, which is nearly bias independent. This is, again, because much of this component is driven by the static spin torque that exists near the boundary of the ferromagnetic layers: this static torque is entirely in the y-direction. We see that $S_y = 0$ for all ϕ and all biases within the S layer. This indicates that excess current in that layer is due to triplet pairs. At high bias, and at all angles ϕ, the S_x and S_z components within the superconductor are quite appreciable. These nonzero spin currents, in S, occur whenever the applied bias is greater than the critical bias (CB). This bias corresponds to an effective value of Δ (smaller than Δ_0): it represents the effective pair potential that the superconductor provides near the F_2/S interface, and has a nonmonotonic angular dependence on ϕ. Thus, the shift in the CB is due to the proximity effect between the F_2 and S layers. The angular dependence comes from the formation of triplet pairs, which is possible only when there is non-collinearity. In this case, with perfect interfaces, the angular dependence of the critical bias is large, confirming previous results for the charge current, see Chap. 8. At $E = 0.6$, the critical bias values for each angle are sometimes above and sometimes below the value of E. For angles such that the CB is greater than the bias ($E = 0.6$ in this case), the excess spin current is zero in the superconductor. However, when the critical bias is lower than the applied bias, the excitations have energy greater than the effective gap energy and at those angles one finds non-zero spin current in S.

In the high bias limit, we see that the spin current is rotated in the $x-z$ plane from the z-axis to an angle close to the mismatch angle ϕ. In the ferromagnetic layers, the spin current precesses about the exchange fields $\mathbf{h}_1$ and $\mathbf{h}_2$ in F_1 and F_2 respectively. The precession in F_2, however, is damped due to the proximity effect of the superconductor, the current becoming constant at the boundary. The spin current in the normal metal layer is also constant, since there are no torques there, and the orientation of the spin current rotates in the $x-z$ plane to an angle between 0 and ϕ, with a nonzero y-component that is due to the net STT in both ferromagnetic layers. One can get a better grasp of the spin current orientation within the system by viewing the spin current in three dimensions, as in Fig. 9.8. There the spin current for $\phi = 90°$ and $E = 2.0$, from the same situation as in Fig. 9.7 is depicted. From this figure, the orientation of the constant spin current in N and S is easier to see, and the precession within the ferromagnets is more readily understood. At high bias, the overall magnitude has a local maximum for $\phi = 90°$.

As to the spin accumulation $\delta\mathbf{m}$, as defined in Eq. (9.11), in the same situation as in the previous figures, its y-component is several orders of magnitude smaller than the other two, for rather obvious reasons, and need not be discussed. The component $\delta m_x(y)$ is zero for $\phi = 0$ and $\phi = 180°$. The component $\delta m_z(y)$ is nonzero and only weakly ϕ dependent in F_1, whereas $\delta m_x(y)$ is oscillatory with y and small in that region. These quantities are also spatially oscillatory in F_2, although larger in magnitude. Furthermore, δm_z and δm_x are nonzero and nearly constant with position in the S region at large bias, but their magnitude decreases slightly as ϕ increases from 0 to 180°. The latter is also generally true in the normal metal region N. In general the behavior of the spin accumulation is oscillatory everywhere at low biases, but with small amplitudes, for this case. It spatially oscillates in N and irregularly rotates in the $x-z$ plane, particularly for mismatch angles near $\phi = 90°$. The overall magnitude increases with bias with unchanging angular dependence. As a vector, it tends to align with $\mathbf{h}_2$ within the superconductor — this is similar to the spin current behavior. The magnitude of $\delta\mathbf{m}$ also decreases as ϕ increases from 0 to 180°.

It is again more illuminating to discuss this quantity in terms of its spatial average than to try to make sense of its oscillations. These averages turn out to be small, because of the oscillating behavior just described, in

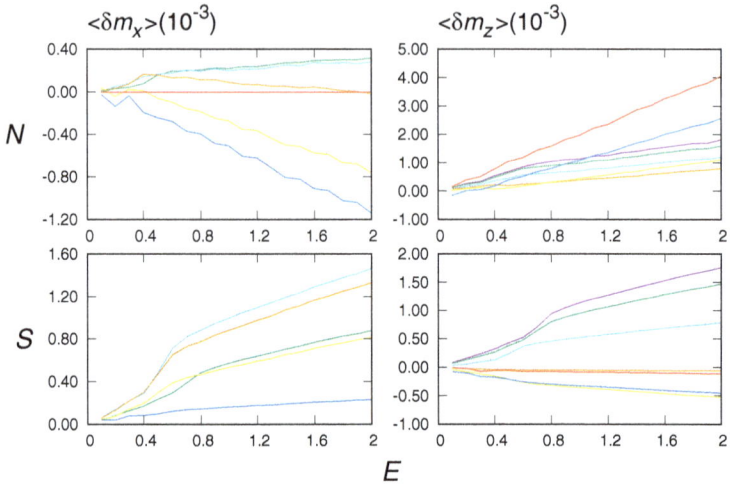

Fig. 9.9 Results for the average x and z components of the spin accumulation (Eq. (9.11)) for the same situation as in Fig. 9.6. In this case, the nontrivial behavior is in the S and N layers. The color key is as in Fig. 9.6.

the F regions. Thus, in Fig. 9.9 there are plots of the spatial average (see Eq. (9.12)) of the spin accumulation only in the S and N superconducting and normal layers, as a function of bias. One can see immediately a critical bias behavior in δm_x, at which value the magnitude rises quickly with bias. Above this value the growth becomes approximately linear. Each component is nonmonotonic with ϕ, and δm_x is maximized between $\phi = 60°$ and $\phi = 90°$. δm_z features a similar, but less dramatic critical bias feature, with this component decreasing for angles $\phi > 90°$.

Much of the above discussion applies to the results for other physical parameter values given below. The above results for ideal interfaces are a basis of comparison as we now introduce realistic interfacial scattering.

9.4.3.2 *Non-ideal interfaces*

What is the influence of interfacial scattering, of a strength realistic for good samples, on the above results? To examine this, we first consider the case where only one barrier exists, at the F_2/S interface, with a rather large scattering parameter value. The layer thicknesses are as in the ideal case of Sec. 9.4.3.1 so that we can focus on the barrier dependence. The interfacial barriers H_B (on either side of the N layer) and H_{B3} (between F_2

and S) are 0 and 0.9 respectively. Recall the remarks below Eq. (8.31) for the notation. When the scattering is relatively large at the F_2/S interface, as in this example, the superconducting proximity effect is reduced. One should compare these results to those found in the zero scattering case of Sec. 9.4.3.1 in order to examine closely how the decrease of the proximity effect influences the spin currents.

In this and subsequent examples we will begin by showing the spin current. Thus, Fig. 9.10 shows results for all components of **S** at two values of E, chosen as in Fig. 9.7. One can see that the x and z components of the spin current are now driven to zero, within numerical precision, at the lower bias value for which results are shown, $E = 0.6$. This is due to the considerable increase in the CB due to the barrier, which makes it more difficult for the triplet states to penetrate through the pair potential in the superconductor. The y component, however, is still nonzero due to the static spin

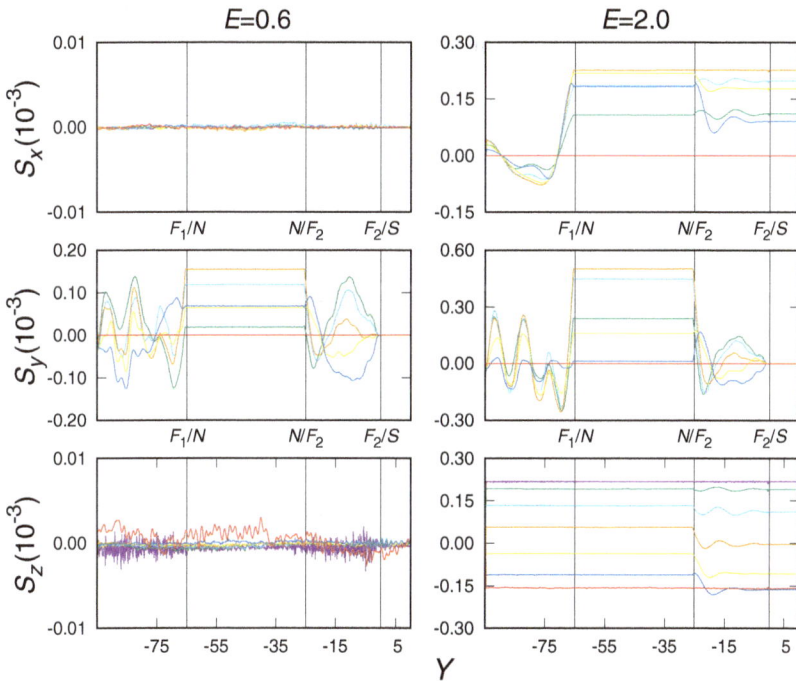

Fig. 9.10 Spin current results with a strong tunneling barrier at the F_2/S interface, at two values of the bias E. The layer thicknesses and other parameters are as in the previous four figures. Results are shown for different values of ϕ, with the key given in Fig. 9.6, repeated below in Fig. 9.12.

torques of the ferromagnetic proximity effect. But, unlike in the previous case discussed, S_y now increases significantly at higher biases, although not as dramatically as the other two components. In the high bias regime, the system returns to precessing about **h** in the ferromagnetic region: **S** rotates about the $x-z$ plane, this time closer to the second ferromagnetic field $\mathbf{h_2}$, which is oriented at an angle ϕ. The overall magnitude of the spin current is of course reduced by the barrier.

As in the ideal interface situation, results for the spin accumulation are best presented in terms of the spatial averages of its x and z components in the N and S layers of the valve. These spatial averages are shown, in the just mentioned regions and for the same barrier values, in Fig. 9.11. As explained in connection with Fig. 9.9 the y component is negligible and the oscillatory behavior in the F layers relatively trivial. As a result of the high F_2/S barrier, the spin accumulation is seen to be significantly decreased in magnitude within S at the low bias limit. However, it increases dramatically at the high bias limit, while remaining smaller than for perfect interfaces. The orientation of the vector $\delta\mathbf{m}$ in S remains fixed to that

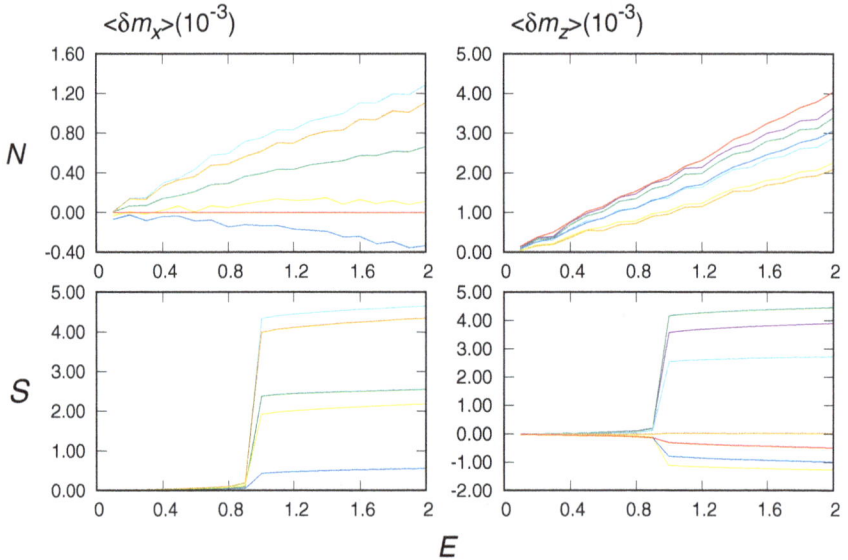

Fig. 9.11 Average x and z components of the spin accumulation with a tunneling barrier at F_2/S interface. Results in both the N and S layers are shown. Geometry and parameters are as in Fig. 9.10. Results are given as a function of bias for several values of ϕ. The key is again as given in Fig. 9.6 or Fig. 9.12.

of the exchange field $\mathbf{h_2}$. In the normal metal, the spin accumulation vector rotates counterclockwise within the $x-z$ plane for $\phi < 90°$ and then reverses direction to become aligned with the z-axis again for $\phi = 180°$. The rotation in the $x-z$ plane is uniform throughout the N layer in the high bias limit, but not in the low bias case. Looking again at the averaged spin accumulation values, a remarkable feature in the superconducting layer, as seen in Fig. 9.11, is a dramatic, sharp increase in the magnitude of $\delta \mathbf{m}$ at the critical bias, after which the magnitude grows at a much slower rate. The angular dependence remains approximately the same as before. The low bias spin accumulation is heavily impeded by the high barrier.

Turning now to the effect of the S/F_2 high barrier on the spin torque, we examine, in Fig. 9.12, the STT results averaged and presented in the same way as in Fig. 9.6. We see that the STT exhibits the same critical bias features seen in Fig. 9.6. However, the high barrier causes the critical bias to increase and to become nearly ϕ independent. Its value is seen to be $E \approx 0.85$ in the results for τ_x (in both F_1 and F_2) and for τ_z in F_2. Furthermore, τ_x in F_2 shifts to become almost entirely negative. The y component is changed dramatically by the barrier: τ_y steadily increases in magnitude with increased bias for all angles except $\phi = 150°$. The static spin torque is heavily impeded by the introduction of a large barrier between F and S, and the ensuing increase in the pair potential at the interface.

Continuing with this study of interfacial barrier effects on spin transport, let us consider now the complementary case where the scattering potentials at both of the F/N interfaces are nonzero, although below the tunneling limit, while the F_2/S interface is ideal. The geometrical and material parameters are of course kept the same as in the previous figures. Specifically the interfacial barriers H_B and H_{B3} (recall Eq. (8.31)) are 0.5 and 0 respectively. Thus, there is a full proximity effect between S and F_2. Perhaps unsurprisingly, the introduction of these barriers turns out to be very important, as the spin-valve effect, which determines much of the spin-transport features, is quite sensitive to these scattering potentials.

We begin again with the spin current. In Fig. 9.13 one sees that it is nonzero in the N region at low bias, as in the ideal case of Fig. 9.7. S_y is now almost entirely bias independent and symmetric within the normal metal layer: it is positive for $\phi > 90°$ and negative for $\phi < 90°$. Similarly, the ϕ dependence of S_x at low bias is nearly symmetrical in all layers. At high bias, the spin current again increases in the x and z component, penetrating the superconductor. Due to the interfacial scattering, the overall magnitude decreases from the zero barrier case, especially for the x and z components.

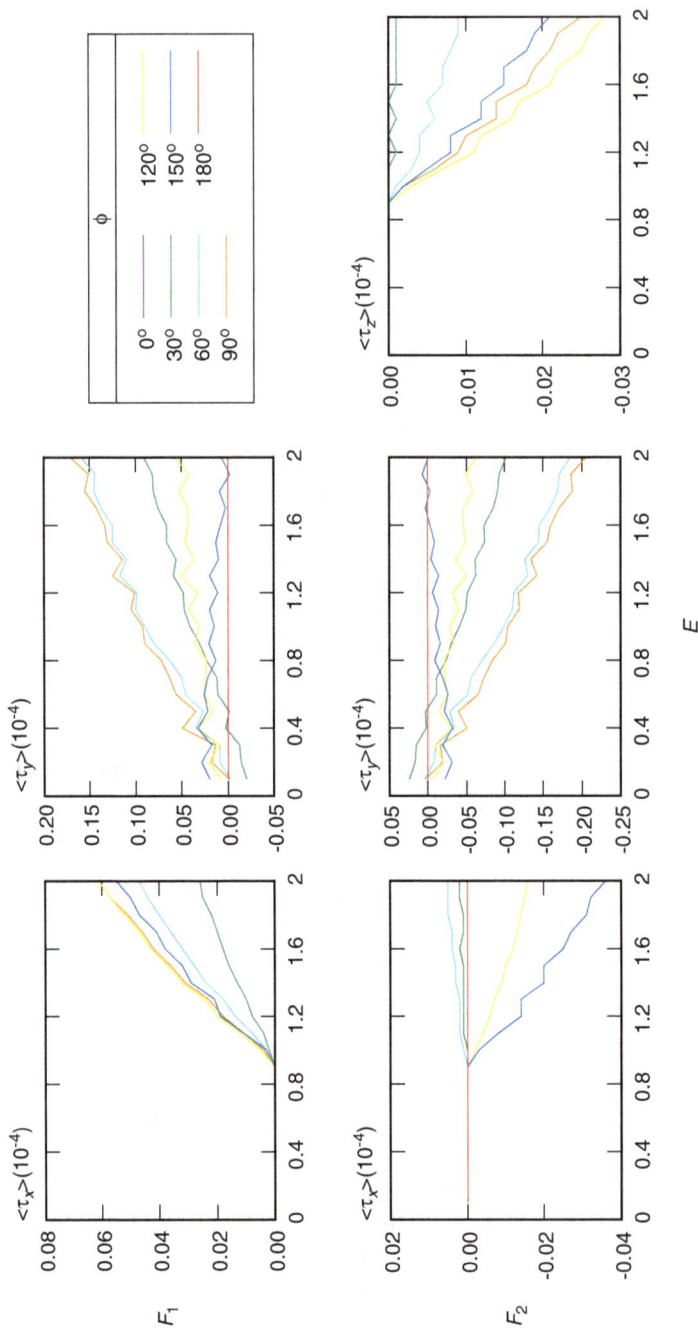

Fig. 9.12   Average spin torque results with tunneling barrier at the F_2/S interface, as in Figs. 9.10 and 9.11. Results are given for all nontrivial components averaged over the relevant regions, and plotted for several values of ϕ as indicated in the key.

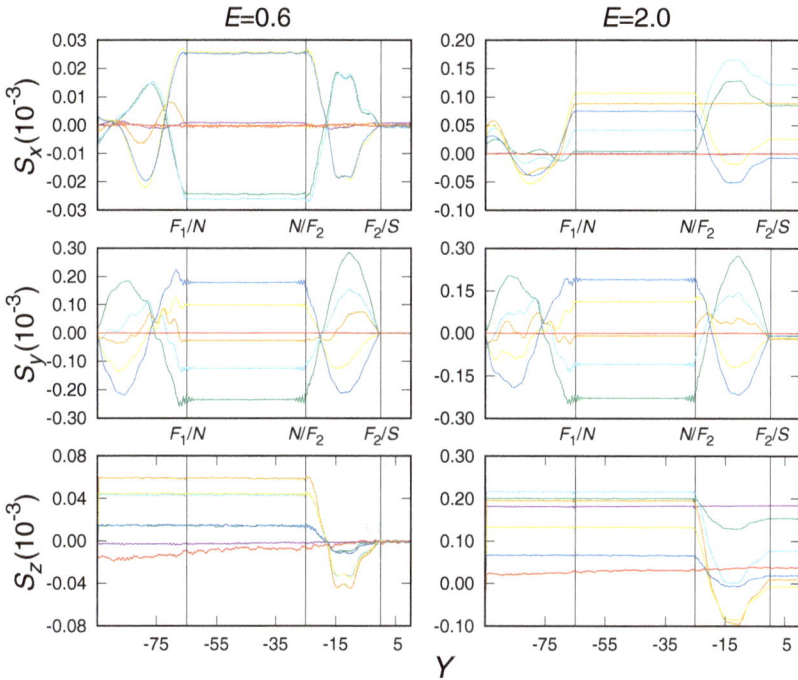

Fig. 9.13 The spin current with nonzero barriers in the spin valve $F_1/N/F_2$ interfaces. The layer thicknesses are as in all other figures in this subsection. The interfacial barriers H_B and H_{B3} are 0.5 and 0 respectively. Results are presented in the same way as in Fig. 9.7. The different curves in the same plot are for different values of ϕ, with the key given as in Fig. 9.12.

For the spin accumulation vector one finds that, in comparison with the ideal interface case, the angular dependence is decreased in the normal metal layer with more oscillations in δm_x about the zero value and a peak forming in δm_z in the low and high bias case. The results are again best presented in terms of the most relevant averages. Thus, in Fig. 9.14 we see an average spin accumulation in S with a similar angular dependence and critical bias features as in the zero barrier case, Fig. 9.9, but with decreased magnitude. One exception is for the x component at $\phi = 150°$, which is significantly larger. In the normal layer, $\langle \delta m_x \rangle$ increases up to a ϕ dependent CB, then steadily decreases for increasing bias. The component δm_z steadily increases at all biases, and has a greater magnitude than δm_x.

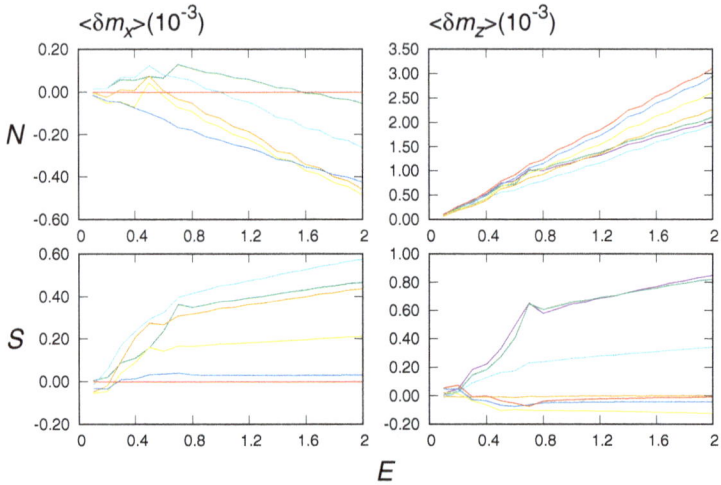

Fig. 9.14 The average spin accumulation in the same spin valve as the previous figure, Fig. 9.13. Results are presented as in Figs. 9.9 and 9.11. The color key is as in Fig. 9.12.

As to the STT, results for its averaged components as a function of bias E, at these barrier values, are given in Fig. 9.15. There, we see significant differences in the behavior of this quantity, as compared to the single high barrier example in Fig. 9.12. The component τ_x in F_1 no longer features any CB behavior: it is nearly constant with E. In both ferromagnets, τ_y is again constant with bias, with a slight increase in the F_1 and a decrease in the F_2 layer. The overall magnitude is significantly smaller, in all layers and for all components, than in the zero barrier case, depicted in Fig. 9.6. In F_2, we see an interesting symmetry of the ϕ dependence emerge in τ_x and τ_z. For τ_x, the values for $\phi = 30°$ and $\phi = 60°$ are both increasing and positive, and correspondingly $\phi = 120°$ and $\phi = 150°$ are decreasing by an equivalent amount. Similarly, for τ_z, we see a similar decrease in value with respect to bias for supplementary angles ($\phi = 30°, 150°$ and $\phi = 60°, 120°$).

To conclude this examination of the barrier dependence of spin transport, we examine the real case where there are scattering barriers at all the interfaces in the spin valve. The layer thicknesses for the F_1NF_2S layers are kept, as before, to $30/40/25/180$ respectively. The interfacial barriers H_B and H_{B3} (see Eq. (8.31)) are set at intermediate and realistic values, namely 0.5 and 0.3 respectively. The spin current is found to be very similar to that shown in Fig. 9.13 and need not be shown again: one can conclude

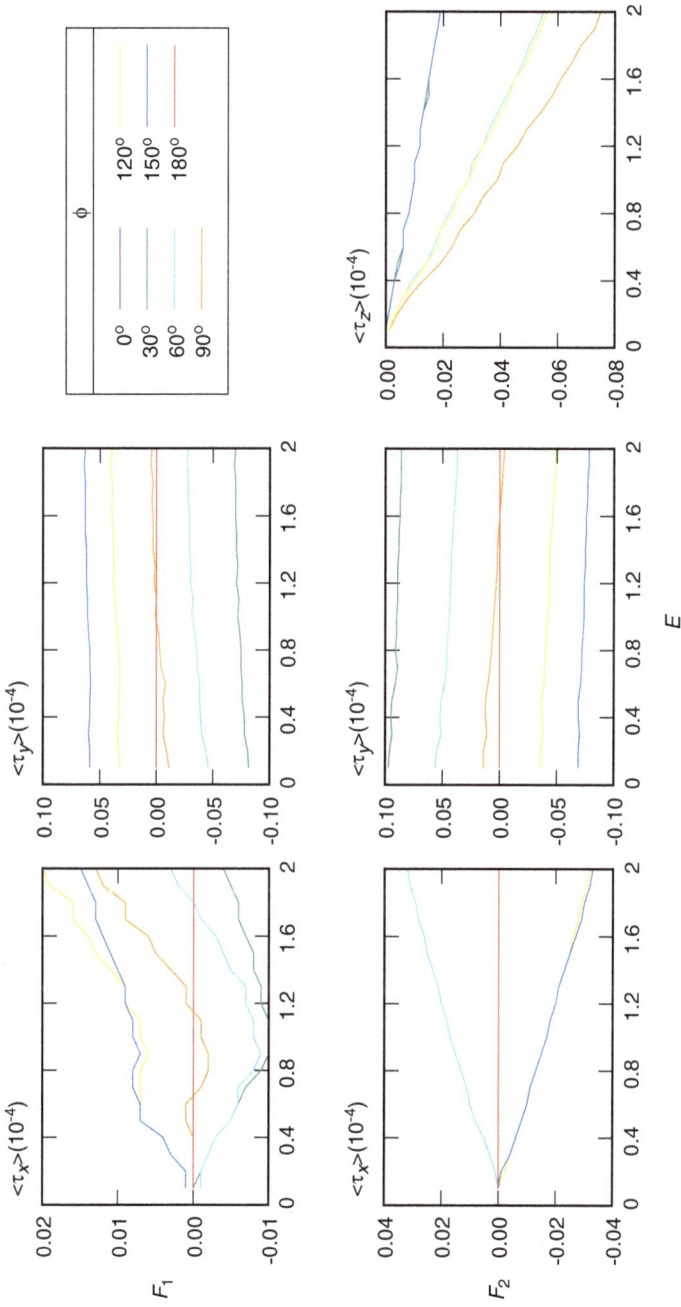

Fig. 9.15 The spin-transfer torque, averaged over the indicated layers and plotted as a function of bias, for the same spin valve as the previous two figures. Results are given for different values of φ as e.g., in Fig. 9.12.

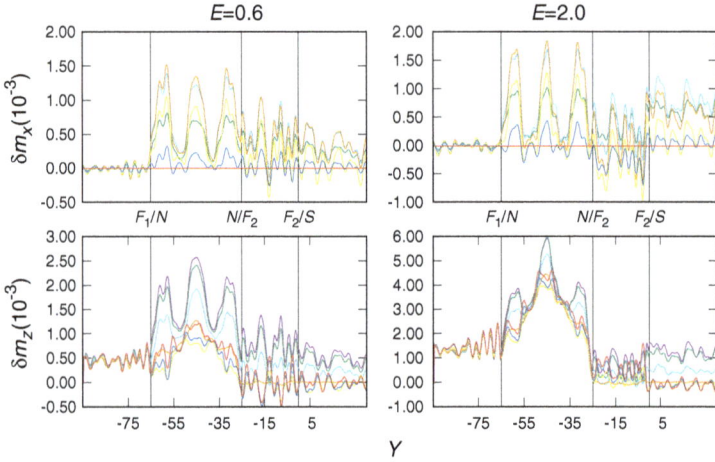

Fig. 9.16 Results for the position dependence of the local spin accumulation with nonzero barriers at all interfaces: the interfacial barriers H_B and H_{B3} are 0.5 and 0.3 respectively. The geometry is as in previous figures. Results are shown at two bias values. Vertical lines locate the interfaces. Only the interesting part of the Y range is included. The different lines correspond to different values of ϕ with the key given in Fig. 9.15.

that the introduction of a moderate size barrier at the F_2/S interface does not significantly impact the spin current. The spin-transfer torques also remain nearly unaffected. These are very gratifying results, from a practical point of view. Their occurrence means that the proximity effect in spin transport is not seriously inhibited by this additional barrier, and the spin-valve effect dominates the spin transport in these cases. Hence I focus here on the spin accumulation, both on its position dependence at fixed bias and on the bias dependence of its average over the relevant layers.

The position dependence of $\delta\mathbf{m}$ is shown in Fig. 9.16. As explained above the y-component is insignificant. Results are seen there for two values of the bias, the larger of which is clearly above the CB. One finds that the vector $\delta\mathbf{m}$ in the normal metal layer departs significantly from what is found in the previous case, at $H_{B3} = 0$: the barriers make a difference. One can observe in δm_z a three peak structure particularly prominent for $\phi < 90°$, although it occurs at all values of ϕ. The x component also forms three peaks at both low and high biases in N, at all angles. As in the previous cases, $\delta\mathbf{m}$ is rotated in the $x-z$ plane in N. These rotations are, as one can see, non-uniform, and are spatially oscillating; the troughs align with

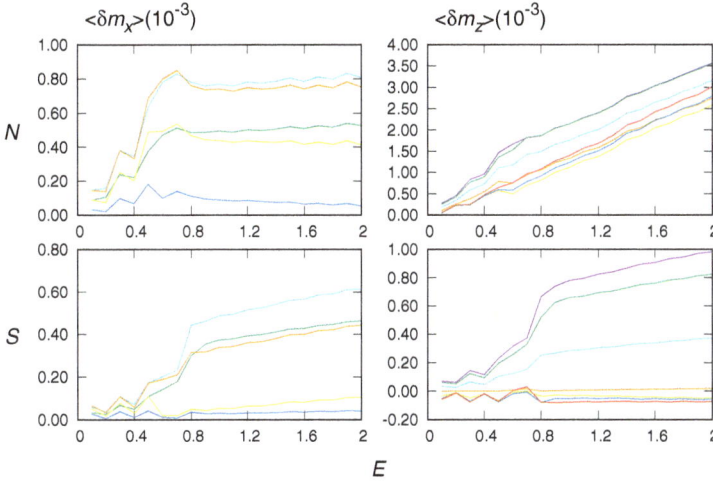

Fig. 9.17 Results for the average spin accumulation with nonzero barriers at all interfaces: the interfacial barriers H_B and H_{B3} are 0.5 and 0.3 respectively. The geometry is as in previous figures. The presentation of the results is as in Fig. 9.9. The color key is as in Fig. 9.15.

the z-axis while the peaks align at an angle smaller than the mismatch angle ϕ.

As to the average spin accumulation results for this case, they are presented in Fig. 9.17, in the same way as previous average spin accumulation figures. One can see an enhancement in the critical bias feature in S seen in Fig. 9.14, reflecting that the low bias conductance is repressed in this case. There is a steep growth in the magnitude of $\delta \mathbf{m}$ at the critical bias. In the normal metal, we see a behavior for the averaged δm_z similar to that in Fig. 9.14 but with a remarkably different angular dependence. For the averaged δm_x in N we see a very different high bias behavior: δm_x now increases dramatically at the critical bias and then abruptly levels off to a slightly decreasing bias dependence. The behavior in δm_x in N is now much more similar to that of the averaged δm_x or δm_z in S.

9.4.4 *Spin in spin valve structures: Effects of geometry*

In this section we turn to the dependence of the spin current results on the geometry (i.e., the combination of layer thicknesses) of the spin valve. The range in which the layer thicknesses can be experimentally varied is rather limited in practice, as discussed in Chap. 6, and our considerations will refer to that range. Consistent with this desire to emphasize experimentally relevant examples, we will stick to the situation where the interfaces are not perfect: the scattering barriers will all be nonzero and in fact, for the purposes of easier comparison, will have the same values as in Fig. 9.17 and the other figures in the last part of Sec. 9.4.3.2, namely $H_B = 0.5$ and $H_{B3} = 0.3$.

In practice, the variation of the S layer thickness is not very relevant: this layer must be thick enough to exceed at the very least the coherence length of the superconductor, and beyond that its precise value is rather unimportant. The spin transport results depend also only weakly on D_{F1}, the thickness of the outer F layer, provided only that this quantity is large enough to efficiently polarize the incoming electrons, as it must be. There-fore, one should focus on varying the intermediate layer thicknesses: that of the normal metal N and of the inner ferromagnet F_2, that is D_N and D_{F2} respectively, and that is what we will do here. The dependence of the results on these two quantities is strong. Specifically we will first look at what happens when the thickness of the the normal metal layer spacer is increased from the previous "default" value $D_N = 40$ of the previous sub-section, Sec. 9.4.3.2, to $D_N = 60$, while keeping the same values for the thicknesses of the F_1 and F_2 layers. Then, the thickness of the inner ferro-magnetic layer will be decreased from the previously studied $D_{F2} = 25$ to $D_{F2} = 15$, while keeping $D_N = 40$. These geometric changes can strongly affect the transmission and reflection amplitudes, as we have seen in the conductance studies in Chap. 8 just as they do in basic quantum mechanics problems. Recall for example the standard homework problem of calculat-ing the transmission across two delta function potentials, where the results can depend drastically on the separation between the two scattering cen-ters. We will see below how these rather minor changes in the geometry affect the spin-transport quantities.

In the first case, that of increasing D_N to $D_N = 60$, the spin current and the spin-transfer torque vary relatively little with bias, so it is appropriate to focus on the spin accumulation. This is done here in Fig. 9.18. The spa-tial structure of this quantity (not shown) exhibits a three peak structure similar although not identical to that which has been mentioned in con-nection with the discussion of Fig. 9.17. The main difference is that the oscillatory structure occurs at all ϕ and all biases. Remarkably, the three peak behavior is inverted in δm_x. Indeed, $\delta \mathbf{m}$ makes a clockwise rotation in the $x-z$ plane in N, counter to both the spin current and spin accumulation

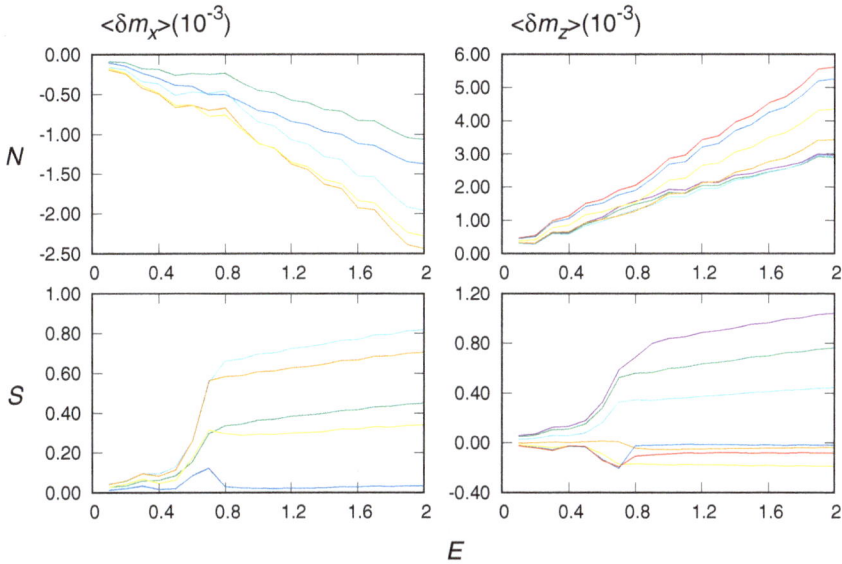

Fig. 9.18 Results for the average spin accumulation in the N and S layers for a spin valve with $D_N = 60$ and the other thicknesses as in previous figures, i.e., 30/60/25/180 for the F_1NF_2S layers respectively. The interfacial barriers H_B and H_{B3} are 0.5 and 0.3 respectively, as in Fig. 9.17. The organization of this figure is also as in Fig. 9.9.

we have seen thus far. Its orientation in S remains unaffected. There is also a significant increase in the magnitude of $\delta\mathbf{m}$ in all layers for high biases, indicating greater growth in the spin accumulation. The behavior of the average spin accumulation, as shown in Fig. 9.18, reflects the comments just made. In the superconductor the results shown for this averaged quantity are similar to those in Fig. 9.17, with increases to the x component for angles $\phi = 30°$, $90°$, and $120°$. The behavior in N is significantly different from that found in the previous cases, where in the x component there was no major critical bias feature and a steadily decreasing bias dependence: this is now similar to the behavior of the z component's magnitude. The z component has the previously noted steady increase with bias, but the angular dependence is now most similar to that in Fig. 9.14. We see then that the ϕ dependence is very sensitive to both the layer thickness and the barriers.

We should turn now to the case of changed D_{F2}. In this situation, as opposed to the discussion relating the D_N, one has to include the results for spin current and torque components because there are nontrivial changes in

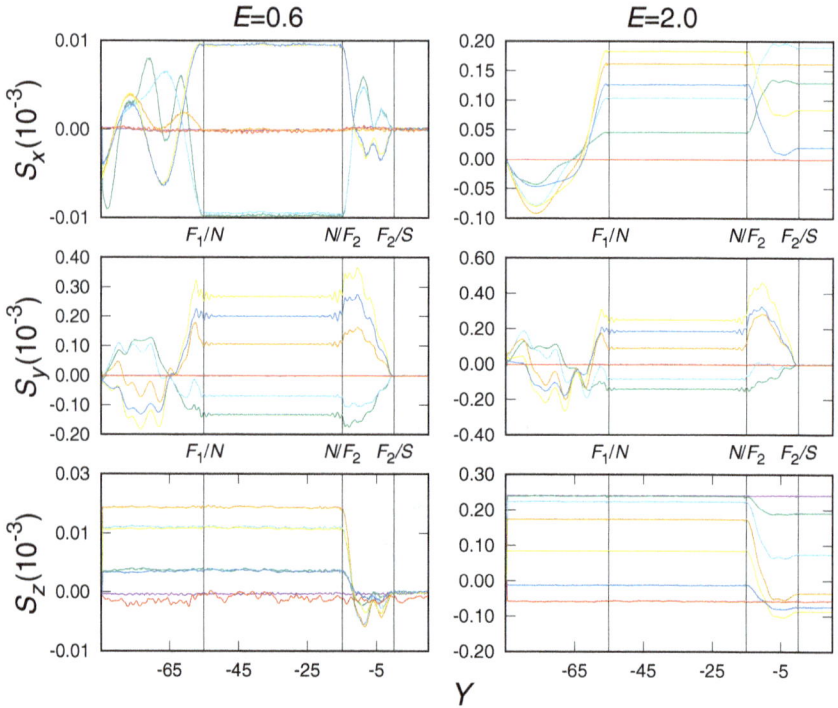

Fig. 9.19   Results for the spin current of the same structure studied in Sec. 9.4.3.2 but with smaller D_{F2}, equal to 15 instead of 25: the layer thicknesses for the F_1NF_2S layers are 30/40/15/180 respectively. The interfacial barriers H_B and H_{B3} are 0.5 and 0.3 respectively, as in Fig. 9.17. The organization of the plots is also as in Fig. 9.13, and the color key is as in Fig. 9.15.

the magnitude and orientation of these quantities. The discussion, therefore, needs to be more detailed. We will decrease D_{F2} to 15, from its
"default" value of 25, and keep everything else, including the realistically
imperfect barrier values as in the last part of Sec. 9.4.3.2.

Let us begin with the spin current. In Fig. 9.19 the results for its components are shown, arranged as in Fig. 9.13. By comparing the results of
these two figures one sees that when decreasing the intermediate ferromagnetic layer spacing from $D_{F2} = 25$ to $D_{F2} = 15$, the x and z components
of the spin current decrease significantly in the low bias limit from their
values at $D_{F2} = 25$. On the other hand, these components increase in the

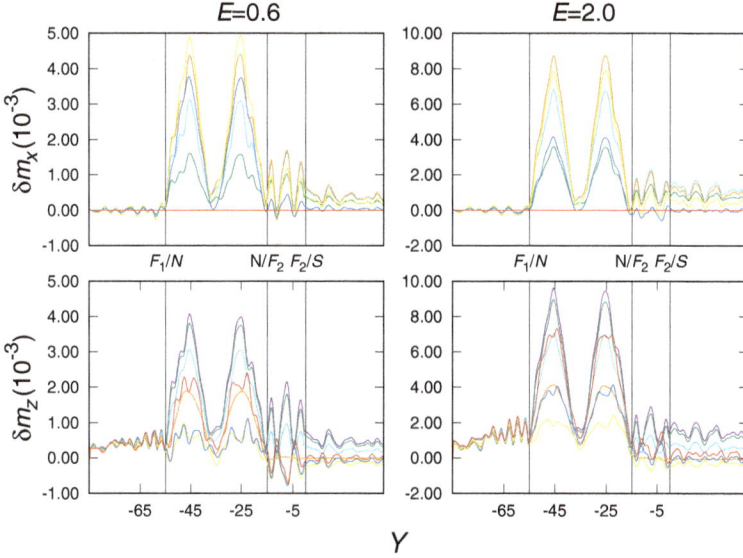

Fig. 9.20 Results for the position dependence of the local spin accumulation with nonzero barriers at all interfaces The geometry is as in the previous figure, i.e., with $D_{F2} = 15$. Results are shown at two bias values. Vertical lines locate the interfaces. Only the interesting part of the Y range is included. The different lines correspond to different values of ϕ with the key given in Fig. 9.15.

high bias limit, particularly the S_x component. The orientation of $\mathbf{S}$ in the superconductor is now rotated closer to the negative z direction, much more significantly for orientations with $\phi > 90°$.

As to the spin accumulation, results are shown in Fig. 9.20. Comparing this figure with Fig. 9.16 one sees at once that decreasing D_{F2} changes the spatial structure of δm_x and δm_z in N from a three-peak to a two-peak structure with the same angular dependence and greater magnitude than at the larger value of D_{F2}. Notice that changing D_N, on the other hand, did not change the spatial structure of the spin accumulation in N, which still had three peaks. Changing the thickness of the inner ferromagnet, changes the rotational behavior of the vector $\delta\mathbf{m}$ in *the normal metal spacer*.

Results for the average spin accumulation are given in Fig. 9.21, displayed in the same way as in Fig. 9.17. The overall magnitude of the spin accumulation components increases dramatically with bias, at a much

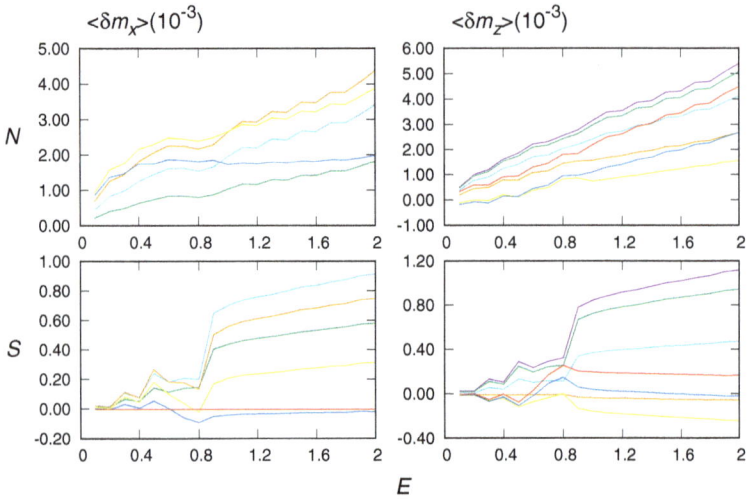

Fig. 9.21 Results for the average spin accumulation, as a function of bias, for exactly the same conditions as in the previous two figures. Results should be compared with those in Fig. 9.17. The color key is again the same as in Fig. 9.15.

greater rate than that seen for the same system at larger D_{F2} or in other systems discussed previously. This can readily be seen by comparing the results in Fig. 9.21 with those in Fig. 9.17. However, $\langle \delta m_x \rangle$ in N steadily increases with bias, with a slight peak near the critical bias. The average spin accumulation at angle $\phi = 150°$ does not increase with bias, and remains a somewhat mysterious outlier.

The results for the averaged spin torque for this structure are shown in Fig. 9.22. Again, the structure of this figure and the meaning of the plots are the same as in Fig. 9.15, and the two figures can be directly compared. A noteworthy feature arises in comparing these STT results with those for the spin current in Fig. 9.19. As remarked in connection with that figure, the orientation of $\mathbf{S}$ in S is in this case rotated to near the $-z$ direction, particularly for $\phi > 90°$. This feature is complemented in Fig. 9.22, where the average spin torque increases its rate of growth. This might seem counter-intuitive at first, but it is important to note that the superconducting pair amplitudes are damped by the ferromagnetic layer.

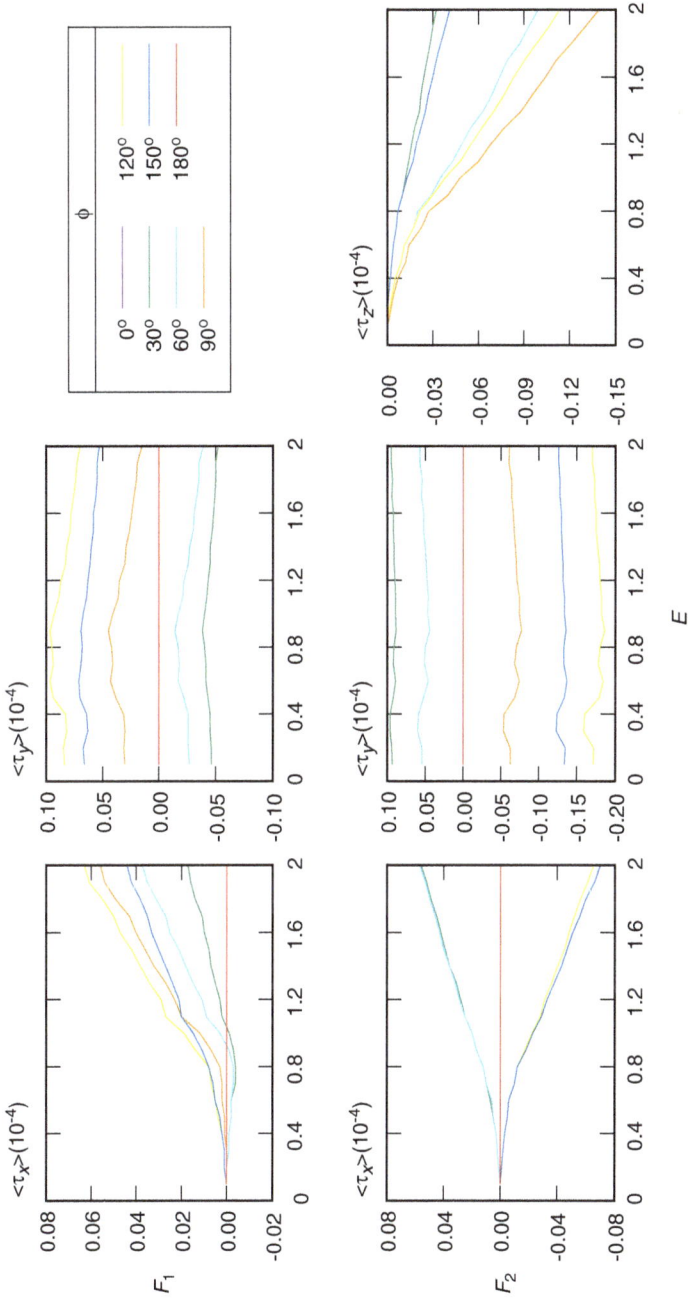

Fig. 9.22 Results for the average spin torque, as a function of bias, in the same situation as Fig. 9.19 and Fig. 9.21. These plots should be compared with those on Fig. 9.15.

9.5 Discussion

In the previous three subsections we have looked in detail to spin transport for $F_1 N F_2 S$ superconducting spin valves. We have seen results for the main relevant spintronic quantities, namely the spin current vector, the spin-transfer torque components, and the local magnetization (or spin accumulation). This has been done for several sets of values of the geometrical parameters of the spin valve, with focus on samples of such thicknesses as can be realistically fabricated, and also including realistic, good but not perfect, interfaces. Material parameters employed, such as internal field and coherence length, have been, as explained in Chap. 6, shown to apply to samples where Nb is the S material, Co the ferromagnet and Cu the N spacer. We have results given as a function of position within the spin valve and of the applied bias, and also spatially averaged quantities. Both the low- and the high-bias regimes, where the bias is smaller or larger than the bulk superconductor pair potential, have been considered. The dependence of all results on the misalignment magnetization angle ϕ between the F layers has been emphasized for two reasons: first, this angle determines the triplet pair formation, and hence the range of the proximity effects and second, as a consequence, it indeed determines the valve action.

The analysis has included a variation of the interfacial scattering parameters and the intermediate layer thicknesses to better encompass a full picture of possible real world results. However, one must keep in mind that the parameter space is exceedingly large and that no naive extrapolations are possible due to the oscillatory behavior of all the quantities involved. Therefore, only a glimpse of the richness and variety of what can be done has been given.

In the results shown one observes a distinctive critical bias behavior where, for a certain (in general ϕ dependent) value of the bias, which is in general somewhat smaller, as we have said, than the bulk S pair potential value, the spin transport behavior changes, with both the spin current and the spin accumulation beginning to penetrate into the superconductor. By looking at plots of the average spin accumulation and STT within each layer, we have seen the CB behavior featured in the magnitude of these quantities. Their spatial averages show distinct growth in the spin accumulation in S, and sometimes also in N for certain sets of both interfacial scattering and thickness parameters. The spin-transfer torque also shares this behavior within the ferromagnetic regions, with an additional symmetrical behavior in the angular dependence when the interfacial barriers are fully introduced.

We have also vividly noted, at higher bias values, the spatial precession of the spin current within the ferromagnets due to the spin-transfer torque. The spin current precesses about the internal field of the ferromagnet, with a spatially decaying amplitude within the intermediate F_2 layer due to the proximity effect of the superconductor. This results in both the spin current and the spin accumulation being oriented within the superconductor at an angle near the field misalignment angle ϕ, and at an angle between zero and ϕ within the normal metal layer. This is only one way in which the misalignment angle plays a factor. Indeed, the critical bias features are angularly dependent chiefly because of the angular dependence of the triplet amplitudes, resulting in a very complex and in general non-monotonic behavior in ϕ for all of the spin transport quantities. We have already seen the angular dependence of the critical bias exhibited in Sec. 9.4.1.

Another noteworthy feature of the spin accumulation occurs within the normal metal layer, where the system transitions, as parameters vary, from a situation where the magnitude of this quantity has a single peak at the center of the normal layer, to multiple peak behavior. By varying *either* the interfacial scattering parameters *or* the normal metal layer thickness, one gets a transition from a two-peak to a three-peak behavior. Naively, one would assume this to be due to the to the normal quantum mechanical effects of the spacial oscillations alone. However, by varying the thickness of the intermediate ferromagnetic layer D_{F2}, a two-peak behavior can occur [32] for the same normal metal layer thickness and interfacial scattering values. This is unique to these spin valve systems, which are highly sensitive to the exact set of parameters, both geometrical and physical. Indeed, the spatial spin current and spin accumulation features can not be extrapolated to trends within the set of parameters we have analyzed. However, the average quantities of the spin accumulation and spin-transfer torque, which have been emphasized, may be at least sometimes extrapolated at high bias values, as the spatial averages tend to be linear in this limit.

We now have spent two lengthy chapters carefully studying, separately, charge and spin transport. But these quantities are coupled: the transport results are related. In the next chapter we will consider the question of how *charge* transport in the different *spin* channels affects the total conductance structure.

Chapter 10

Spin and Charge Transport: The Voltage Dependence of the Conductance

In the previous two chapters, Chaps. 8 and 9 we have studied charge and spin transport, respectively. Although we have mentioned that the two are intimately related, we have treated them separately. Not surprisingly, this leads to a somewhat incomplete picture of the overall phenomenology. Indeed, many features of the dependence of the conductance G on the bias E remained unexplained. We will now see in the present chapter that a reexamination of how charge is transported in the spin up and down channels leads to a much better understanding of the physics involved in these features.

When considering the charge transport in Chap. 8 we learned that the conductance $G(E)$ versus bias voltage V curves in $F_1/N/F_2/S$ valve structures (see Sec. 8.4.4) can exhibit a "subgap" (E below the CB) peak structure. In this chapter we will see that in general the low bias structure of G in spin valve devices is due to what we will call "spin split conductance" behavior: in spin valves charge is transported differently in the up and down spin channels. We will study in some detail the features involved and what parameters influence them. The spin dependence of the conductance can lead to exotic behavior and unusual properties, e.g., in the layer thickness dependence in such structures. By studying the spin-split conductance, we will gain a deeper understanding of the full conductance features seen in Chap. 9. In addition to seeing the results of exact numerical calculations, an approximate but very simple analytic model of an NFS structure with infinitely thick N and S layers will be used for pedagogical purposes.

The results of this approximation help illuminate the physics and in gaining a better understanding of the fully self-consistent numerical calculation for a finite nanoscale system.

10.1 Definitions and General Considerations

The structure numerically considered in this chapter is then the spin valve structure in Fig. 2.1. For the NFS structure that will be studied analytically, one has to imagine that both the N and the S layer are very thick, while the leftmost (F_1) layer is absent. For the numerical calculations, the same parameter value choices, based on experimental considerations, as in Sec. 9.4.2 will be made, both as geometrical and material (Co, Nb) parameters, and as to interfacial scattering.

Let us carefully define what do we mean by the separate spin conductances, or conductance in separate spin channels. Following the expression Eq. (8.21) for the total conductance, we can see that we should define $G_\uparrow$ and $G_\downarrow$ via:

$$G(\epsilon) = \sum_{\sigma=\uparrow\downarrow} P_\sigma G_\sigma(\epsilon) \tag{10.1}$$

$$= \sum_{\sigma=\uparrow\downarrow} P_\sigma \left(1 + \frac{k_{\uparrow 1}^-}{k_{\sigma 1}^+}|a_{\uparrow,\sigma}|^2 + \frac{k_{\downarrow 1}^-}{k_{\sigma 1}^+}|a_{\downarrow,\sigma}|^2 - \frac{k_{\uparrow 1}^+}{k_{\sigma 1}^+}|b_{\uparrow,\sigma}|^2 - \frac{k_{\downarrow 1}^+}{k_{\sigma 1}^+}|b_{\downarrow,\sigma}|^2 \right),$$

where, as you can see from Eq. (8.21), we are considering the forward conductance, setting $\theta_i = 0$. This is the quantity that we call the "conductance" (or sometimes the "weighted conductance") by default in this and other chapters. The weight factors P_σ are defined in the last sentence of the paragraph following Eq. (8.21). They are not part of the definition of G_σ.

Generally, $G_\uparrow$ and $G_\downarrow$ will differ significantly. However they are related to each other by a rotation in spin space around the y-axis. The rotation is performed by the unitary transformation Eq. (6.3) which can be alternatively written as $U = e^{-\frac{i}{2}\theta\sigma_y}$, where θ is the angle of rotation (not to be confused with the angle of incidence which was set to zero near the end of the previous paragraph!). Taking then the expectation value, one can redefine the spin-up and spin-down conductances, $G_\sigma(\theta)$, in a basis rotated from that of the z-axis, $G_\sigma(0)$, as:

$$G_\uparrow(\theta) = \cos^2(\theta/2)G_\uparrow(0) + \sin^2(\theta/2)G_\downarrow(0) \tag{10.2a}$$

$$G_\downarrow(\theta) = \sin^2(\theta/2)G_\uparrow(0) + \cos^2(\theta/2)G_\downarrow(0) \tag{10.2b}$$

In an NFS system the angle θ can be thought of as the angle ϕ between the field in F and the z-axis, since this basis rotation is exactly the same as a rotation in F. This is *not* the case in the $F_1 N F_2 S$ spin valve when there is an actual angular mismatch between the internal fields of the two F layers. It will prove useful to compare the change in the spin split conductance due to the angular mismatch to that arising from pure basis rotation.

10.2   Spin Split Conductance Results

We go now to the results. The figures in this section are all from Ref. [33].

This section consists of two parts, the first analytic, the second numerical. In the first, we will consider a simple NFS system, with infinitely thick S and N layers. It can be solved analytically in a non-self-consistent approximation, that is, by setting the pair potential to a constant Δ_0 in S. Now, I have repeatedly said that non-self-consistent approximations are very incorrect. But in this case, as we will see, there are very useful qualitative lessons one can learn, and it is always more illuminating to see results emerge analytically, rather than from a computer file. The analytic calculations for this rough model are discussed below. The results, although quantitatively inaccurate, qualitatively illustrate important physical points that apply to all F/S systems. The rest of the results, in the second subsection, will use the proper self-consistent approach. It will become clear overall that the behavior of the spin split conductance as one changes the intermediate ferromagnetic layer thickness or the interfacial scattering barriers, particularly that at the N/F interface, is particularly interesting. The plots are chiefly for the same, experimentally motivated, parameter values used in Sec. 8.4.4, as mentioned in the previous subsection, except that (to clarify some important physical issues) the value of h will be allowed to roam a bit more. The dimensionless coherence length is $\Xi_0 = 115$ and D_S is of course always substantially larger.

10.2.1   *Analytic results*

If one relinquishes (temporarily of course) self-consistency, one can derive expressions for the relevant amplitudes which are in principle analytic. For an NFS heterostructure where N and S are of infinite thickness, but D_F is finite, the expressions for the incident waves, impinging from N, are of the form given in Eqs. (8.1) and (8.2) but with a simplified wavevector structure involving, instead of e.g., Eq. (8.3) only the spin independent wavevectors

$$k_N^\pm = \left[1 \pm \epsilon - k_\perp^2\right]^{1/2}. \tag{10.3}$$

For the intermediate layers, the eigenfunctions contain both left- and right-moving plane waves, as we saw in Sec. 8.3.2. Thus the wavefunction for the intermediate F layer has eight unknown coefficients. The S layer contains right-moving quasiparticles and left-moving quasiholes, with four unknown coefficients. Plane wave expressions for Ψ_F and Ψ_S are as in Eq. (8.10) (with $\phi = 0$) and Eq. (8.7).

The continuity condition at each interface $\Psi_N(0) = \Psi_F(0)$, $\Psi_F(D_F) = \Psi_S(D_F)$, where for this infinite system, one can choose the N/F interface to be at $Y = 0$. The conditions on their derivatives are $\partial\Psi_N(0)/\partial Y = \partial\Psi_F(0)/\partial Y + 2H_B\Psi_F(0)$ and similarly for the second interface. The transfer matrix method of Sec. 8.3.2 can then be used to write these as 8×8 matrices $\mathcal{M}_i$ multiplied by their respective vector of unknown coefficients x_i for each layer i, as explained in Chap. 8. Then, $\mathcal{M}_N x_{N,\sigma} = \mathcal{M}_{F,l} x_F$ and $\mathcal{M}_{F,r} x_F = \mathcal{M}_S x_S$, where (l, r) denote that the wavefunctions are evaluated on the left or right side of the layer respectively and σ the spin of the incoming electron in the N layer. By solving and eliminating the intermediate layer coefficients, one finds the eight total coefficients of both the N and S layer:

$$x_{N,\sigma} = \mathcal{M}^{-1}{}_N \mathcal{M}_{F,l} \mathcal{M}^{-1}{}_{F,r} \mathcal{M}_S x_S. \qquad (10.4)$$

Solving these eight equations simultaneously for both spin-up and spin-down incoming electrons, the two sets of four reflection amplitudes $b_{\sigma,\sigma'}$ and $a_{\sigma,\sigma'}$, one set for each incoming spin state σ', are found.

Formally this calculation is analytic, although the full form solution for each reflection amplitude can not be written in a compact manner. However, knowing the form of the plane wave description does help in understanding the spatial dependence of the amplitudes. This comes from a combination of plane waves in F, of the form $e^{ik_\sigma^\pm d_F}$, in which the wavevectors in the F layer are defined by Eq. (8.3).

To begin the discussion of the results of this introductory example, let us examine Fig. 10.1. There we see simple plots of the spin-split conductance G_σ (i.e., the spin-up and spin-down components) and the total weighted conductance, G, as defined by Eq. (10.1). In this case, because of the absence of an outer F layer, the weight factors P_σ are simply $1/2$. These conductances are plotted as functions of applied bias. The results plotted there are for a moderately strong barrier, $H_B = 0.5$, located at the N/F boundary. As we saw in Sec. 8.4.5, a small amount of interfacial scattering allows for the formation of the subgap conductance peak. This peak occurs for specific thicknesses of F at biases between zero and the critical

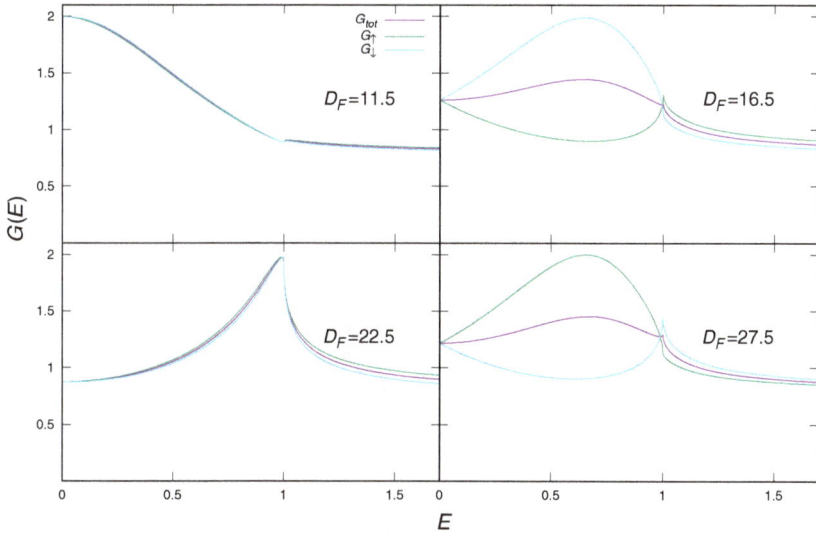

Fig. 10.1 Spin-up and spin-down contributions to the total weighted conductance, $(G = (1/2)(G_\uparrow + G_\downarrow)$, also plotted, see label in the first panel) as a function of bias E in an infinite NFS system. A single barrier with $H_B = 0.5$ is at the N/F interface. The internal field in F is $h = 0.145$. The four panels are for different values of the F layer thickness, D_F, chosen at intervals of a quarter period ($\pi/2$ phase) of the spatial dependence (see text). These are approximate analytic results. Notice the behaviors of $G_\uparrow$ and $G_\downarrow$ reverse between the top and bottom panels on the right. In the text, panels are referred to as panel one (upper left), panel two (upper right), panel three (lower left) and panel four (lower right).

bias (the "subgap" bias region). As we shall see, the spin-split conductance components can differ dramatically from each other, especially near the CB. In this single F layer system, the total G is simply the average of the up and down spin-band contributions. When the spin-up and spin-down conductances are split from each other, a peak in the total conductance occurs at the bias where they differ the most.

Examining the peak value of the conductance in Fig. 10.1 , we can see a periodic behavior with D_F, with a periodicity of $\pi/h \approx 22$ in dimensionless units, since the value $h = 0.145$ found experimentally appropriate for Co in Chap. 6 is used. This periodicity can be traced, of course, to the well-known oscillations of the Cooper pair amplitudes, reflected in the above given value, as has been discussed at the end of Chap. 2. Figure 10.1 includes four panels, each for a different value of D_F within one cycle of this periodic behavior. In the first (upper left) and third (lower left) panels, which correspond to a D_F difference of about half a period, the peak in the total G occurs at zero bias and at the CB respectively, while in the similarly separated second (upper right) and fourth (lower right) panels there is a subgap bias conductance peak. The total conductance peak moves away from zero bias in the first panel to a finite subgap bias value in the second. Increasing D_F further, the peak moves to the critical bias in panel three, then returns in panel four to the same subgap bias value as in panel two. It goes back to zero bias, with a peak feature very similar to that in panel one for $D_F = 33$ (not shown): at that point a whole period in D_F has elapsed. In the first and third panels there is little difference between the subgap spin-up and spin-down conductances. On the other hand, there is a very large difference in the spin-split conductance for the second and fourth panels. In the second panel, the spin-down conductance has a large subgap peak, with G reaching a value of $G = 2$ before decreasing towards the CB, where there is a discontinuous change in slope (leading to a "shoulder"). The spin-up conductance has the opposite behavior, with a dip in the subgap region that increases to a sharp cusp shaped peak at the CB. This spin-split conductance then yields a total G with a local maximum at the spin-down conductance's maximum, which is also the spin-up conductance's minimum. In panel four, one can see a very similar situation but with the behaviors of $G_\uparrow$ and $G_\downarrow$ *reversed*. $G_\uparrow$ has now an intermediate maximum and a shoulder CB feature, and $G_\downarrow$ an intermediate minimum and cusp critical bias feature. In both of these panels, the CBC is also split between spin-up and spin-down, and the total G has a hybrid cusp-like behavior. There is then a crossover value, where each component (and the total conductance) meet, at a bias slightly below that of the CB.

In all four panels the ZBC is the same for the spin-up and spin-down conductances, and consequentially for the total weighted G. In ordinary Andreev reflection, a spin-up electron reflects into a spin-down hole, and vice versa. In the zero bias limit the electron and hole have equal energy. Thus, in this single F layer case, the zero bias spin-up transmission amplitudes are the same as those for spin-down, due to the symmetry of the electron/hole traveling in the spin-up/spin-down bands. We will see below that this is not the case when there is a second ferromagnetic layer.

Specific details of the D_F periodicity structure can be clearly seen in Figs. 10.2 and 10.3. In Fig. 10.2 the ZBC is shown as a function of D_F for $H_B = 0.5$ (top panel) and $H_B = 0.9$ (bottom panel) at $h = 0.145$, and also at $h = 0.0725$ and $h = 0.29$, half and double the original value. These choices best demonstrate the dependence of the spatial periodicity

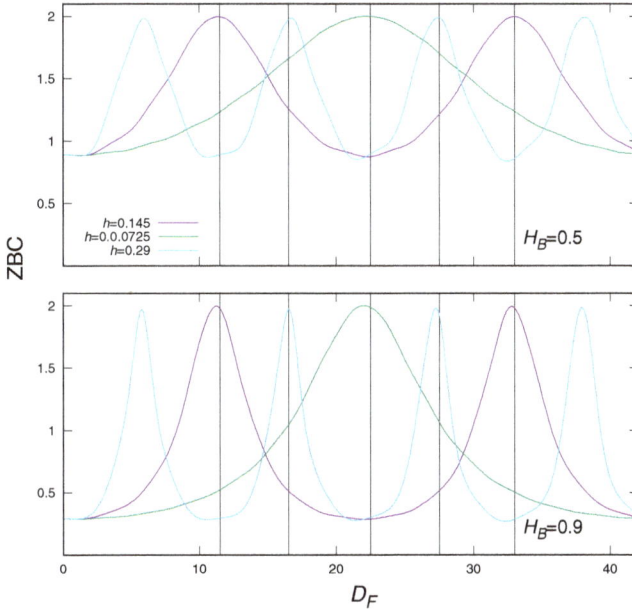

Fig. 10.2   The zero bias conductance (ZBC) for an infinite NFS system, plotted vs. D_F, for values of $h = 0.145$ (curves with two peaks, top and bottom panels), $h = 0.145/2$ (one peak curves) and $h = 2 \times 0.145$ (one peak). The top and bottom panels correspond to $H_B = 0.5$ and $H_B = 0.9$ respectively, at the N/F interface. The D_F range includes about two oscillation periods at $h = 0.145$. The wavelength is π/h in all cases.

on h. As mentioned above, the ZBC is equal for the spin-up, spin-down, and total conductances and therefore only the total weighted G is shown. The four leftmost vertical lines in each plot are the values of D_F used in Fig. 10.1 and the fifth is for $D_F = 33$ at which value one full cycle is complete for $h = 0.145$. One can clearly see here the π/h dependence of the wavelength of the oscillation. For a value double the original, the wavelength is halved, and vice versa. The oscillatory behavior looks very regular and fairly sinusoidal at $H_B = 0.5$, except for some minor irregular variations which are more prominent for $h = 0.29$. However, for the larger barrier value of the bottom panel, the oscillatory pattern is less sinusoidal, with a sharper dependence of the ZBC on D_F at the ZBC maxima and a broadening of the ZBC minima. Near the vertical lines, there is a slight change in the phase of the oscillation for the stronger barrier. The periodic behavior breaks down for values of D_F of less than a quarter-period, where the ZBC becomes constant and independent on h.

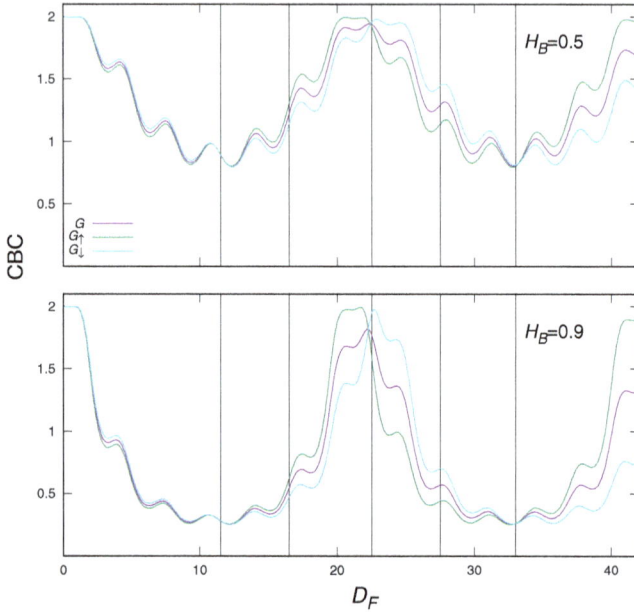

Fig. 10.3   The critical bias conductance (CBC) (weighted total, spin-up, and spin-down) for an infinite NFS system vs. D_F. The top panel and bottom panels have $H_B = 0.5$ and $H_B = 0.9$ respectively, both at the N/F interface. The D_F range is about two periods of the π/h oscillation. There are also smaller, superimposed oscillations with periodicity π.

Following up on this, in Fig. 10.3, the CBC is shown as a function of D_F, for both $H_B = 0.5$ and $H_B = 0.9$ (top and bottom panels respectively) at $h = 0.145$. This is done for the spin-split conductance components, which do not have the same CBC value, as well as for the total weighted conductance. The same overall periodic structure as in the ZBC occurs, with a π phase difference, since the CBC maxima occurs at the ZBC minima. There is also a minor oscillatory behavior with wavelength π superimposed on the broader π/h oscillations: this is unobservable in the ZBC. The spin-up and spin-down conductances cross over at the CBC maxima and they also converge at the CBC minima (where there are ZBC maxima). Between nodal points there is a difference in the spin-split conductance components that reverses between a dominant spin-up or dominant spin-down conductance. The separation becomes greater as D_F increases, or as the barrier strength increases.

Therefore, these approximate analytic results show a regular, periodic behavior in the conductance features as a function of the ferromagnetic layer thickness. There is also a subgap bias conductance peak, the prominence of which increases with the strength of the scattering barrier at the N/F interface. This peak is due to the splitting of the spin-up and spin-down conductances. This analysis will be helpful in interpreting the numerical results below.

10.2.2 *Numerical results*

Let us turn now to the real world of self-consistent numerical results.

To make the discussion of these numerical spin valve results more under-standable, it is useful to include also a brief discussion of a simpler finite size layered NFS structure, with a single barrier, with $H_B = 0.5$ at the N/F interface: this is somewhat similar to the case studied in the ana-lytic results. The calculation is now numerical, three-dimensional, and fully self-consistent.

In Fig. 10.4, the total conductance G and its spin up and spin down components (Eq. (10.1)) are plotted vs. the scaled bias voltage E. The four panels corresponding to varying intermediate F layer thickness D_F, with fixed $D_N = 90$ and $D_S = 180 = 1.56\Xi_0$. The scattering strength at the N/F interface is $H_B = 0.5$. The variation in D_F is chosen, as in the non-self-consistent results of Fig. 10.1, to include a thickness variation that encompasses a full period of the conductance's subgap peak periodic in D_F behavior. The most obvious difference between the results of the non-self-consistent analytic calculation in Fig. 10.1 and those in Fig. 10.4, obtained via the numerical self-consistent procedure is that the latter case leads to a varying critical bias. This is directly related to the drop in the pair potential due to the proximity effect of the pair amplitude.

The first (upper left) panel of Fig. 10.4, corresponds to the situation where the ZBC is large and the CBC is low. The critical bias itself is sig-nificantly smaller in the self-consistent case than in Fig. 10.1. There is little difference between the spin-up and spin-down conductance curves. Just as in the analytic case, this behavior is periodic with D_F. In the second (upper right) panel, the transition in the spin-split conductance becomes appar-ent, with a subgap peak in the total G due to the opposing behavior of the spin-up and spin-down components: $G_\uparrow$ displays a positive concavity and a cusp feature at the critical bias, while $G_\downarrow$ shows a negative concavity with a weak shoulder feature at the CB, similar to those found in the analytic

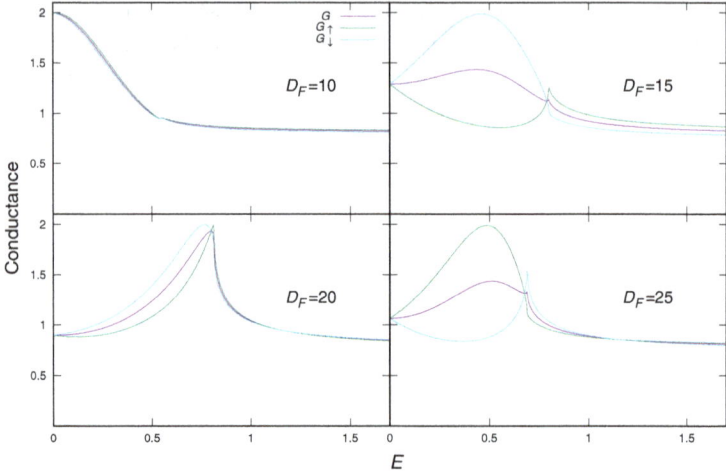

Fig. 10.4 Spin-up, spin-down, and total weighted G (see Eq. (10.1)) vs. bias E in a lay-ered NFS system with $D_N = 90$ and $D_S = 180$. Conductances are calculated numeri-cally and self-consistently. A scattering barrier is present at the N/F interface. The four panels are for different values of D_F, chosen at intervals of a quarter period of the spatial dependence. This figure should be compared with Fig. 10.1. Notice that the behaviors of $G_\uparrow$ and $G_\downarrow$ reverse again between the top and bottom right panels. References to panels in the text are as mentioned in the Fig. 10.1 caption.

calculation. Although the CB conductance depends on the spin, the CB value itself of course does not. This is because both spin channels interact with the same effective pair potential, which for the single-ferromagnet sys-tem, and singlet pairing, is spin independent. In the third (lower left) panel, the spin-split and total conductance peak locations are seen to converge towards the critical bias. Although not shown here, the relative behavior of the spin-up and spin-down conductance switches abruptly for slightly different values of D_F, with a sharp transition similar to what is seen in the ZBC peaks of Fig. 10.1 in the analytic calculation. In the fourth (lower right) panel, there is another subgap conductance peak similar to that in the second panel but now with the spin-split conductance components switch-ing behavior, as was the case in the analytic calculation. The spatial period considered in the four panels corresponds to a wavelength π/h for the cho-sen value of h. We conclude that the self-consistent behavior of the $N/F/S$ conductance qualitatively displays the same periodic behavior as revealed by the analytic non-self-consistent calculations. However, the CB is now dependent on D_F as can clearly be seen scanning carefully the four panels.

10.2.2.1 *Spin valves*

Now, let's move on to the full-fledged spin valve. This is the case of major theoretical and practical interest. There the behavior is more complex, since the dependence on the field misalignment angle ϕ affects both the spin-split conductance peaks *and* the critical bias. The ϕ dependence is particularly important when applied to spin valves, as any orientational dependence in the conductance constitutes a "valve effect" that can be exploited. We have seen such ϕ effects in G in Chap. 8, e.g., in Secs. 8.4.4–8.4.6 and also in the previous chapter for spin transport (e.g., Secs. 9.4.3.2 and 9.4.4). Here I continue to focus, as in the just mentioned previous chapters, on the intermediate F_2 layer dependence and the oscillatory behavior of the peak conductance, as already noted in the NFS case. Therefore, we keep D_{F1} and D_N fixed (at values 30 and 60 respectively) and vary D_{F2} over a moderate range of values encompassing a full period, as explained above. The results shown are for a small subset of interfacial scattering values, and focus on the spin-split effects that arise as D_{F2} varies.

In Figs. 10.5–10.7 there are, respectively, plots of $G_\uparrow$ $G_\downarrow$ and the total weighted G as a function of voltage for several values of the misalignment angle ϕ. The spin valve structure considered has $D_{F1} = 30, D_N = 60, D_S = 180$, and material parameters as in previous figures. In this case, there is a single interfacial barrier, located at the N/F_2 interface and having the moderate value $H_B = 0.5$, so that it can be compared with the analytic approximate results in Fig. 10.1 or the numerical ones in Fig. 10.4. This barrier configuration best introduces the behaviors of the peak conductance that can occur. Below, we will see results including a full set of barriers. In each of these figures the panels correspond to different values of D_{F2}, as indicated. In these figures, we see that the spin-up and spin-down components (Figs. 10.5 and 10.6 respectively) are highly dependent on the relative angle of magnetization. It is obviously no coincidence that the spin-up conductance very closely resembles that of the spin-down conductance for supplementary values of the angle ϕ. Much of this resemblance is due to the change in ϕ being accounted for, in large part, by a purely mathematical rotation of the spin-split conductance as given by Eq. (10.2). From that equation it is seen that under a rotation by an angle θ, $G_\uparrow(\theta) = G_\downarrow(\pi - \theta)$ and vice versa. The angular dependence of each spin component closely resembles a combination of $\phi = 0$ of the spin-up and spin-down conductance, rotated into the respective ϕ basis via Eq. (10.2) for $\theta \to \phi$. For the same reason, it should be no surprise that a subgap peak in the total spin-split conductance, Fig. 10.7, is found near $\phi = 90°$, since this can be largely described by a combination of the spin-up and spin-down conductances, as is the case with the total conductance. However, not all the differences in the features between $G_\uparrow$ and $G_\downarrow$ can be explained by this rotation: an intrinsic ϕ dependence exists that is different for each component of the spin-split conductance. This yields a much more complex angular dependence in the total conductance (Fig. 10.7). Let us now see how this works.

In the $G_\uparrow$ plots of Fig. 10.5, one can see a considerable spread in the critical bias. The dependence on ϕ is relatively weak in the first panel and becomes much stronger in the other three. In the second panel, the CB increases for angles greater than 90° and decreases for angles less than 90°. In the third and fourth panels (bottom row), we see the opposite: the CB decreases for $\phi > 90°$ and increases for $\phi < 90°$. Recall that in the NFS case we saw the spin-up and spin-down conductance swap behavior in panels two and four of Fig. 10.1, with a transitional behavior shown in panels one and three. Similarly, one sees here the CB behavior also making this transition in its ϕ dependence. The cusp and shoulder behavior of the CBC is not qualitatively changed by the introduction of the outer ferromagnet. We also see a split in the ZBC. One should notice that, although in the second panel the value of the quantity plotted in the prominent $E \approx 0.5$ region is largest in the antiparallel configuration and smallest in the parallel, $\phi = 0$ case, the ϕ dependence is not quite monotonic. The situation reverses in the fourth panel, but the lack of monotonicity remains.

The results for $G_\downarrow$ in Fig. 10.6, show that the behavior of this quantity is similar to that of the spin-up conductance, Fig. 10.5, but with a mismatch angle dependence shifted approximately by π. The ϕ dependence of the ZBC and the CBC have dramatically changed, with opposite behavior. The ZBC no longer has a crossover in the low bias regime. In effect, the introduction of the outer ferromagnet takes the crossover node of the ZBC seen in the NFS case (see Fig. 10.4) and moves it to the right in the spin-up conductance and to the left in the spin-down conductance. The CBC for the spin-down conductance experiences broadening, in direct opposition of the spin-up conductance, as can be seen best by comparing the right hand panels (panels two and four, right side) of Figs. 10.5 and 10.6. The mismatch angle dependence of the CB also broadens in these two panels. In the lower left panel (panel three), we see the CB values move closer together and reverse the order of their angular dependence. This is explained by the spin-up conductance and spin-down conductance being at different phases

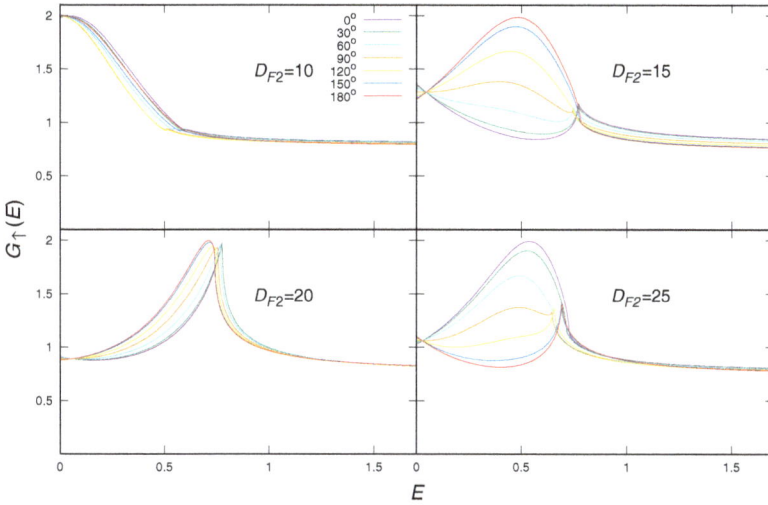

Fig. 10.5 Spin up conductance vs. bias for the values of ϕ indicated in the legend. The plots are for an $F_1 N F_2 S$ spin valve structure with $D_{F1} = 30, D_N = 60, D_S = 180$ and a barrier of strength $H_B = 0.5$ at the N/F_2 interface. The four panels are for different values of the intermediate F_2 layer thickness at intervals of one quarter of its period variation. At $E = 0.5$ on the upper right panel, the curve for $\phi = 180°$ is at the top, and that for $\phi = 0$ at the bottom, while in the lower right panel they are reversed. References to panels in the text for this and all subsequent figures in this chapter are as described in the Fig. 10.1 caption.

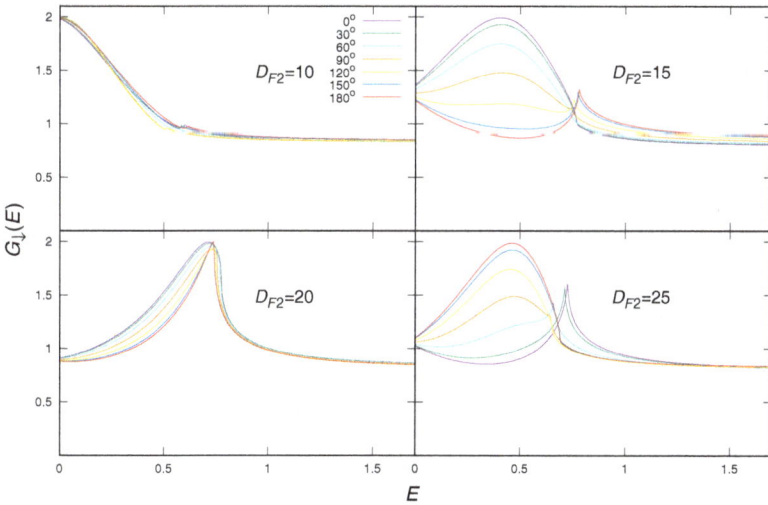

Fig. 10.6 Spin down conductance vs. bias for several values of ϕ, for the same example as in Fig. 10.5 and depicted in the same way. See text for discussion.

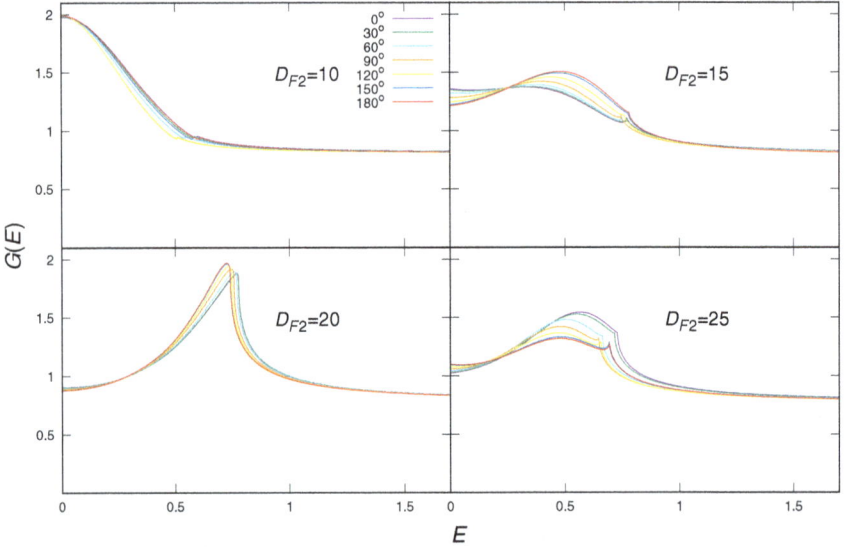

Fig. 10.7   Total weighted conductance vs. bias, for several ϕ, as in Figs. 10.5 and 10.6.

in their D_F periodicity. In panel three, we see the critical bias (and the overall conductance) behavior transition in its angular dependence from that of panel two to that of panel four. From the CB features plotted, we see, due to the spin-valve effect, that the spin-down conductance is slightly advanced in its D_{F2} dependence, while the spin-up conductance lags behind.

Finally, by looking at Fig. 10.7, we can analyze the overall impact of the outer ferromagnetic layer by plotting the total (weighted) conductance, which can then be compared to that in Fig. 10.4. The total G, as given by Eq. (10.1), is not, unlike in the NFS case, simply the average of the spin-up and spin-down conductances, because the outer electrode F_1 is populated with a majority of spin-up electrons: the total conductance is now weighted more heavily towards the spin-up value. Therefore one finds an angular dependence in the total conductance that is more reminiscent of that of the spin-up contribution. This can be seen in the similar CB angular dependence as well as the ZBC and CBC dependences. The combination of spin-up and spin-down conductance leaves us with a smaller subgap peak in the total G, for all angles. Generally, we see a significantly reduced angular dependence when compared to the spin-split conductance, except for the ZBC and the CBC. We also see that the cusp and shoulder CB features are less pronounced. There is a crossover node in panels two and four (right panels) as we saw with the spin-up conductance — however this is not an exact "node"" as the conductance does not cross over at precisely the same bias for all angles. In panel three, we actually see a peak conductance monotonically increasing with ϕ, even though neither the spin-up nor spin-down conductance feature this monotonic behavior. In this transition, the phase difference of the spin-up and spin-down oscillations with respect to D_{F2} has a greater impact on the CBC behavior of the total G than for thicknesses such as those in panels two and four.

If one increases the single barrier strength parameter from $H_{B2} = 0.5$ to $H_{B2} = 0.9$, which decreases the overall proximity effects, there are some changes in the results, [33]. The approximate supplementary angle relation in the ϕ behavior of the spin-split conductance remains. The angular dependence in the CB and the ZBC, seen in Figs. 10.5 and 10.6, becomes much smaller: this reflects the overall suppression of the proximity effects by the higher barrier. Furthermore, the difference between the spin-up and spin-down conductances (besides the switching of conductance behavior to supplementary angles) is clearly diminished. There is, however, a small broadening in the angular dependence of the CB, as well as a better defined cross-over node. This leads to a total conductance that has, in the zero bias and critical bias regions, little angular dependence. However, the subgap conductance peak still maintains a strong angular dependence, rivaling that of the $H_{B2} = 0.5$ case. This is because much of the angular dependence comes from the difference in spin-up and spin-down electron populations emanating from the F_1 layer, in which the large difference between spin-up and spin-down conductance counteracts the decrease in angular dependence of the other conductance features. Indeed, for higher barriers this subgap peak is more pronounced. The reason is twofold: the increased difference between the spin-up and spin-down conductances creates a large peak in the total G, and the decrease in the angular dependence of the CB provides less overlap, which prevents the hybridizing of the cusp and shoulder CB behaviors, and makes the drop-off sharper at the CB.

We finally get to where we want to: the realistic case of a spin valve where there are realistic barriers at all three interfaces. As an example, let us consider $H_{B1} = H_{B2} = 0.5$, $H_{B3} = 0.3$ for the F_1/N, N/F_2, and F_2/S interfaces respectively. As repeatedly mentioned above, and also emphasized in Chap. 6, these values are in the appropriate range for actual experimental conditions. Results are plotted in Figs. 10.8–10.10.

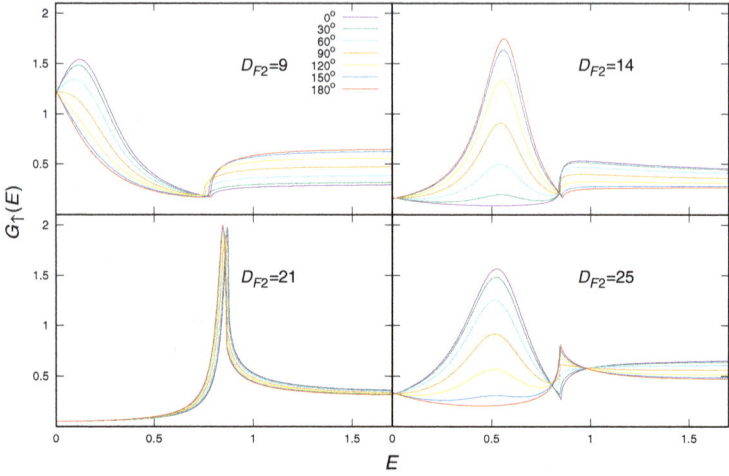

Fig. 10.8 Spin up conductance vs. bias, at several values of ϕ, as indicated. The plots are for a $F_1 N F_2 S$ spin valve with thicknesses $D_{F1} = 30, D_N = 60, D_S = 180$ and realistic barrier values H_{Bi} equal to 0.5, 0.5, and 0.3 at the F_1/N, N/F_2, and F_2/S interfaces respectively. The four panels in each subfigure are for different values of the intermediate F_2 layer thickness in the intervals of one quarter period in the D_{F2} oscillations.

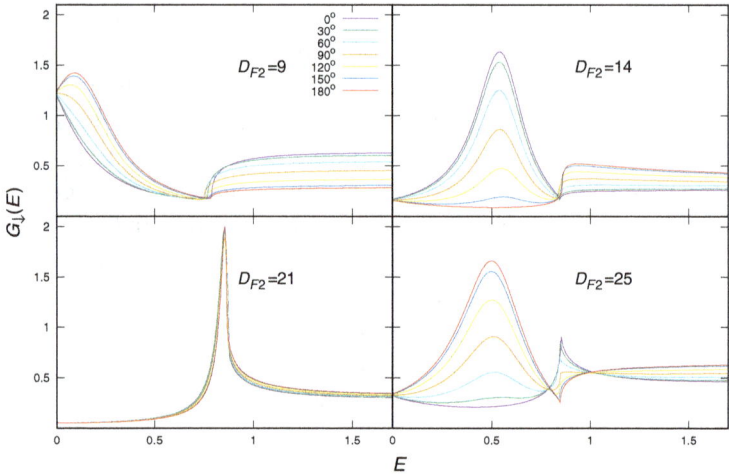

Fig. 10.9 Spin down conductance vs. bias at several ϕ values, under the same conditions as in Fig. 10.8.

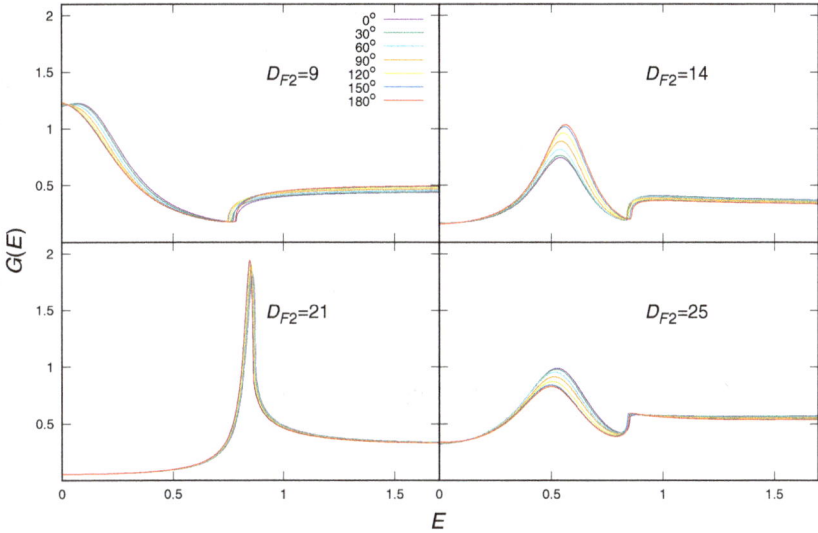

Fig. 10.10   Total conductance vs. bias and several ϕ values, as in Figs. 10.8 and 10.9.

The introduction of the F_2/S barrier can be expected to slightly flatten the subgap peak conductance feature because a barrier at that interface reduces the proximity effect. However, this effect remains small, and when coupled with the moderate F/N barriers, it leaves the conductance with well defined peaks, as we will see in Fig. 10.10 below. Having two interfacial barriers, particularly with similar values, can produce resonance effects in the low bias conductance for certain thicknesses as we saw already, for example, in Sec. 8.4.5. There are a large number of parameter choices that can affect the conductance features in a variety of ways, but what is of interest here is how robust the oscillatory subgap peak behavior is.

In Fig. 10.8 we can see plots of $G_\uparrow$ for this case, while in Fig. 10.9 there is $G_\downarrow$. These plots are arranged as in Figs. 10.5 and 10.6. The first thing to observe is the slight change in the D_{F2} values displayed in the first three panels from the values used in the previous figures just mentioned. In the first (top left) panels of these two figures, the thickness value shown is that closest to the transition where the subgap conductance peak is now closest to zero bias. The ZBC peak conductance values in these panels are very sensitive to small changes in D_{F2}, and the choice of D_{F2} in the plots is tuned precisely to that value. This sensitivity indicates that these barriers can have a large impact on the phase of the oscillatory behavior of the peak conductance. In this plot one observes a large angular dependence at the peak bias, which is at E slightly greater than zero, but hardly any angular dependence on the ZBC. Also, the angular dependence in the subgap bias range is large. This is found also in the single barrier case. The high bias conductance ($E > 1$) now displays a large angular dependence in panels one, two and four, but this dependence is much smaller in panel three (lower left) where the subgap peak transitions to the CB. Examining the spin-up and spin-down conductances, we see considerable broadening with ϕ in the CB and the CBC, in panels two and four. In the first and second panels, we see some slight phase advancing or lagging, but the other conductance features are quite similar. These transitional panels have peaks which are sharper than in the previous results, and the ZBC peak is lower. In comparing the ϕ dependence in thse two figures, the previously made remarks on supplementary angles obviously apply.

Now, in Fig. 10.10 the total conductance for this example is shown. Despite the F_2/S barrier, the subgap conductance peak, as seen in panels two and four, is quite sharp. Furthermore there is a noticeably large angular dependence in the CB, as well as in the high-bias conductance. The ZBC, however, has a smaller angular dependence, but does still feature a small-bias crossover point before the subgap peak conductance. In panel three we see again an angular dependence in the CBC similar to that in Fig. 10.7. Overall, the salient point is that the subgap peak behavior is not only present, but in fact quite pronounced, with a large angular dependence in the peak conductance in panels two and four. This peak conductance is oscillatory with D_{F2}, with only a slight change in phase resulting from the introduction of realistic barriers.

This robust angular dependence of the peak conductance, that is, this spin valve effect, can potentially be exploited, as the subgap conductance peak leads to an angularly dependent change in the excess currents at high biases.

10.3 Summary

In this chapter we have analyzed the spin-split conductance in F_1NF_2S spin valve structures, with a view to gain some physical understanding of the peak and periodicity structure of the $G(V)$ results of Chap. 8. The results presented for these spin valves were obtained via numerical, self-consistent methods. Some results have also been presented for NFS systems using an approximate but analytic method. The main plots assume imperfect interfaces and experimentally relevant parameters as discussed in Chap. 6. There is a peak in the subgap conductance that is periodic with the intermediate F layer thickness. This peak conductance is due to the separate behavior of the spin-up and spin-down contributions to the total conductance, which we have called the spin-split conductance. The subgap conductance peak position oscillates between the zero bias and the critical bias values as D_{F2} varies. At least one spin band conductance has a maximum not far from $G = 2$ at a single bias value in the subgap region — near that bias value the opposite spin band has a minimum. At this subgap bias, there is a pronounced peak in the total conductance due to the spin-up and spin-down conductances being very different at this intermediate subgap bias range, while they converge in the ZBC and CBC. When there are N/S interfaces with moderate tunneling barriers, a peak in the conductance occurs at the critical bias before decreasing to normal conductance [25]. With F/S interfacial scattering there is a subgap peak that is robust to interfacial scattering. It has a large angular dependence in the F_1NF_2S spin valve.

These spin valve results are numerical. It is often difficult to gain physical intuition from purely numerical results. In an effort to gain additional intuitive understanding, we have in this chapter also shown results which were obtained via an approximate, non-self-consistent, analytic approach for an infinite NFS structure. We have uncovered and clarified the physical origin of the spatial periodic behavior by examining the ZBC and CBC as a function of the thickness of the F layer. In both cases, the periodic spatial dependence is due to the interaction of the spin dependent plane wave amplitudes in the ferromagnet, which leads to a wavelength of π/h in the conductance peak (h is the exchange field of the intermediate ferromagnet). In the CBC there are also smaller, superimposed oscillations of the spin-split conductance that switch dominance across the regular nodal points at the CBC extrema. The location of the subgap conductance peak oscillates between zero bias and the CB. The spin-split conductance as a function of bias consequently switches behavior between spin components with changes in F layer thickness, transitioning across the CBC and ZBC peak conductances. The subgap conductance peak is only weakly dependent on the barrier strength. Furthermore, when this peak occurs in the middle of the subgap region, the ZBC and CBC values decrease at a faster rate with increasing barrier strength, leaving a more pronounced subgap peak at higher barrier strengths.

Considering the numerical self-consistent results for the three dimensional $F_1 N F_2 S$ spin valve, one finds the same periodic effects, with an additional dependence of the spin-split conductance on the CB. This is also found for a finite NFS structure. This dependence on the CB is reduced by high barriers: very high barriers kill all proximity effects.

We have looked also at the dependence of the spin-split conductance on the angle ϕ between the internal exchange fields in the magnets: the quintessential valve effect. In the finite NFS system, this amounts to a purely mathematical rotation of our basis. For the $F_1 N F_2 S$ valve system this is of course not the case. Only part of the angular dependence of the spin-split conductance in this system can be attributed to rotations in spin space (see Eq. (10.2)) but since (except at $\phi = 0$ and $\phi = \pi$), S_z does not commute with the Hamiltonian, the dependence on ϕ is beyond that arising from a choice of spin quantization axis. This affects the CB, the CBC, and the ZBC in different ways for the spin-up and spin-down components, causing broadening of the CBC peaks and CB values, as well as shifting the cross-over points where G is approximately equal for all angles ϕ. From the analysis performed in Sec. 10.2.2.1 one can conclude that the phase of the periodic D_{F2} dependence is angularly dependent for both components of the spin-split conductance: the thickness at which the G behavior transitions depends on ϕ. There is a general shift in the spin-split conductance's bias dependence, in opposite directions for each component, with a nodal point, located at zero bias in the NFS system, shifting to higher bias values for spin-up and to lower ones for spin-down. The end result is that the total conductance has a complex angular dependence, where the subgap peak becomes less prominent. This is because the relative shift of the spin-split conductance implies that each component's respective (at supplementary angles) extrema are no longer aligned, leaving their combination (i.e., the total conductance) more smeared, and the other conductance features less pronounced. Nevertheless, a subgap conductance peak with a very strong angular dependence occurs in the $F_1 N F_2 S$ structure. This angular dependence is protected by the subgap peak, which does not diminish strongly with increasing barriers.

The subgap conductance peak, due to the spin-split conductance, is an important and prominent feature that one hopes can be exploited in future superconducting spintronic devices. Considerations similar to those developed in this chapter should help determine and improve the efficacy of a superconducting spin valve in which the valve effect is defined by the angular dependence of the exchange fields. The sub-gap peak is well defined when the interface between the superconductor and the valve is reasonably clean, even when the interfacial scattering within the valve is far from negligible. Although this can lead to very low angular dependence when the peak conductance is at zero bias or at the CB, the angular dependence is large and robust against interfacial scattering for definite values of the intermediate ferromagnetic layer thickness. By tuning the thickness to one of these intermediate values, a valve effect in the excess current can be attained, as a very large angular dependence in the spin-split and total conductance then occurs. This would have a considerable effect on the quality of such spin valve devices.

Chapter 11

Ferromagnetic Josephson Junctions: Statics

11.1 Introduction

In previous chapters we have studied the static and transport properties of ferromagnet/superconductor nanostructures of the spin valve type. Now, in this chapter and also the next, we will consider ferromagnetic Josephson junctions, in which an additional superconductor layer is present next to the F_1 layer, which was previously the outer layer, so that one has a combination of N and F layers, such as an FNF or an FFF trilayer, sandwiched between two superconducting layers.

First, we will study in this chapter the equilibrium properties of these junctions, focusing on the current/phase relationship, which, we will see is not simply sinusoidal, as it is in elementary [26] situations. Then, in Chap. 12 we will learn how to calculate certain transport properties, specifically the quasiparticle conductance as a function of bias voltage.

The basic phenomenon of the current-phase relationship is described as the DC Josephson effect [3, 26]: when a phase difference, $\Delta\varphi$, exists between two superconductors (S) separated by a non-superconducting material the corresponding charge supercurrent is directly controllable via $\Delta\varphi$. The interplay between ferromagnetism and superconductivity implies that junctions that contain one or more central ferromagnetic (F) regions will have properties quite different from those of standard junctions. These properties will provide an avenue for the study of spin currents that can be manipulated. We will see that novel and interesting phenomena arise. Moreover, if the non-superconducting region includes two or more F layers with unaligned magnetizations (e.g., a $SFFS$ structure), manipulation of the angle between the magnetization vectors will, as we have seen in other contexts, lead to long range triplet supercurrents. Additional control of the magnetic state can also be available from the spatially varying spin current within the F layers of the junction, causing mutual torques, as discussed in Chap. 9, to act on their respective magnetic moments. Therefore, we will see that $S(F)S$-based junctions that contain multiple F layers (as symbolized by the parenthetical (F) notation) offer various opportunities for controlling the charge and spin currents, and their influence on the magnetization in terms of the torques the multiple layers induce.

It is of fundamental importance to understand the basic physics of the interplay between spin and charge currents and the long-range triplet pairing associated with the supercurrents. For these purposes we consider in this chapter nanoscale $S(F)S$ Josephson junctions, where the (F) region may, for example, include three layers, FNS, namely two metallic ferromagnets separated by a non-magnetic normal metal spacer leading to an $SFNFS$ spin valve-like structure. Or, it may comprise three F layers, as in five-layer $SFFFS$ junctions. A Josephson supercurrent in these structures is of course established via a phase difference $\Delta\varphi$ between the S terminals.

We have considered examples of the generation of long-range triplet proximity effects in superconducting heterostructures with magnetic inhomogeneity in Chaps. 4 and 5. Triplet Cooper pairs with same-spin electrons are not subject, as this book has repeatedly emphasized, to paramagnetic pair-breaking and can propagate for large distances inside the ferromagnet. Such equal-spin triplet correlations thus play an important role in transport, as we have already seen. Now, we shall see that the oscillatory and long-range pair correlations lead to new behaviors in the current-phase relations (CPRs) in $S(F)S$ with a nontrivial magnetic structure.

11.2   Basic Ideas and Conservation Laws

To make the different configurations studied clear, we display in Fig. 11.1 a generic example of the most general structure considered in this chapter. This is clearly more general than that shown in Fig. 2.2 or the SFS structures whose thermodynamics we have studied in Chap. 4. Figure 11.1 and the other figures in this chapter are extracted from Ref. [34]. The systems studied are particular examples of that depicted in Fig. 11.1. For example, there may be fewer layers. Or, in some cases the middle F_2 layer will actually be an N layer, making a spin valve structure, or it may even be absent. Please note carefully the definition of the angles and that, in this

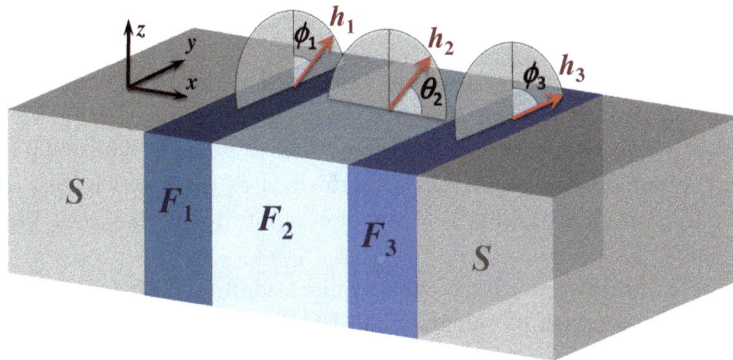

Fig. 11.1 Schematic example of a $SF_1F_2F_3S$ Josephson junction as considered in this chapter. A generic configuration is shown, described by the in-plane magnetization angles, ϕ_i, and out-of-plane angles, θ_i ($i = 1, 2, 3$). The ferromagnetic exchange fields $\mathbf{h}_i$ in each region are expressed as $\mathbf{h}_i = h_i(\cos\theta_i, \sin\theta_i \sin\phi_i, \sin\theta_i \cos\phi_i)$.

chapter, it is the x-axis (not the y-axis) what marks the direction normal to the layers. Do not confuse the angles ϕ_i with the phase difference $\Delta\varphi$. Note also that the angles θ_i are taken in terms of the direction normal to the layers, i.e., the x-axis, while the angles ϕ_i are in the $z-y$ plane, which is the plane of the layers. Thus, we are writing for the internal fields:

$$\mathbf{h}_i = h(\cos\theta_i\hat{x} + \sin\theta_i \sin\phi_i\hat{y} + \sin\theta_i \cos\phi_i\hat{z}), \qquad (11.1)$$

where $i = 1, 2, 3$ denotes one of the magnetic layers. We will again make the realistic assumption that the magnitude of the exchange field is the same in all magnetic layers. However, the angles θ_i and ϕ_i will in general be taken to vary from layer to layer. Also, again in this chapter only, the fields will be occasionally allowed to have a component normal to the layers. Demagnetization effects usually suppress this component but there is no harm in including it occasionally.

The self-consistent diagonalization methods used in this chapter are exactly the same as previously discussed in Chap. 2, see Sec. 2.3 through Sec. 2.5, in particular Sec. 2.5.1. The generic form of the Hamiltonian is the same (see Eq. (2.1)). The presence of the extra S layer poses of course no serious difficulty: it affects mainly issues of what we may call "bookkeeping" in keeping track of the spatial dependence of coupling constants and the internal fields. Some matters of detail are different: because of the phase difference, the pair potential cannot be everywhere real. Hence one must choose as the basis on which to expand the eigenfunctions (the set $\phi_q(x)$

in Eq. (2.14)) as plane waves rather than sine functions. This is of course not a problem. Because of the larger number of layers, the required matrix elements (which of course are now calculated in terms of exponentials) are more intricate than, for instance, those given in Eq. (2.19), but still they involve only elementary integrations. The presence of additional layers makes the numerical computations, performed as discussed in Sec. 2.5.2, technically more demanding in terms of computational resources, but not unduly so.

There is, however, an issue that requires additional attention and due care. Because we wish to calculate the DC Josephson current in terms of the phase difference $\Delta\varphi$ we need a means of imposing this phase difference, as a boundary condition, as it were. It is necessary to ensure that the phase difference between the right and left ends of the sample remains $\Delta\varphi$. The iterative procedure used to reach self-consistency, as discussed in Sec. 2.5 particularly in Sec. 2.5.2, involves an initial guess and subsequent iteration. In this Josephson case, one initially assumes a constant amplitude form for the pair potential in each S layer, but with a total phase difference $\Delta\varphi$ ($\varphi = 0, \varphi = \Delta\varphi$ at each S region). One then finds the quasiparticle energies and amplitudes by diagonalizing the resultant matrix. From them, an iterated pair potential $\Delta(x)$ is found and, by repeating the procedure, it is eventually self-consistently determined through the entire region. But it is clear that the phase itself will change, as well as the amplitude, during the iteration process. This would not necessarily keep $\Delta\varphi$ fixed. In fact, considerations related to free energy minima (Chap. 4) indicate that it would iterate to 0 or π as in Sec. 4.3. To prevent this, in the calculations one assumes that the S layers are relatively thick: eight times the correlation length is assumed in this chapter. Then, for a small region near the sample edges the pair potential is fixed, after each iteration, to its bulk absolute value, with fixed phases $0, \Delta\varphi$ at each end. This is repeated iteratively until convergence is achieved. When determining the current-phase relations, $\Delta\varphi$ is defined as the difference in phases between the outermost parts of the superconductor S layers. As self-consistency evolves with each iteration, the final $\Delta\varphi$ will in general differ slightly, in the S regions away from the edges, from the fixed difference $\Delta\varphi$ that is set in the edge regions. Even in the edge region the value after the last iteration may be different from the prescribed $\Delta\varphi$. Thus, to have the final $\Delta\varphi$ fixed to a prescribed value while varying any other parameters, such as orientation angles or thicknesses, additional calculations may be needed with slightly different initial choices for the phase difference $\Delta\varphi$. As seen from the discussion in

Sec. 7.2, when current is flowing through the junction then in the regions where the pair potential is calculated self-consistently the charge current is spatially constant in the steady state. The edge regions, where the phase difference is fixed, provide the physically necessary source or sink of current, via the applied electrodes, thus acting as an effective boundary condition.

11.2.1 *Conservation laws*

Once again we have to be extremely careful with the conservation laws. This is done as in Sec. 7.2, but there are some minor differences. One is the rather trivial (although possibly annoying) question of having, in this chapter, notation where the x-axis is normal to the layers. Note, however, that this affects spatial labels: the spin axis of quantization remains the z-axis, which is in the plane of the layers. But a separate and more substantial issue is that now we do **not** have an applied voltage driving the currents, but a phase difference. Hence we have to be careful with some of the arguments used in Chap. 7, which may not apply. The most important of these is that, in discussing Eq. (7.8), we dropped the first term in the middle line of that equation because, we argued, it would be zero when the pair potential is real. This was correct then, but it is not now because of the phase difference: the pair potential is not real. On the other hand, the current does flow at zero voltage, when the term in the last line of Eq. (7.8) does vanish. So, at zero temperature and zero voltage the current is exactly the term we dropped in deriving Eq. (7.8), namely:

$$j_x(x) = -\frac{e}{m} \sum_{n,\sigma} \text{Im}\left[v_{n\sigma} \frac{\partial v_{n\sigma}^*}{\partial x}\right]. \tag{11.2}$$

where we have taken into account, in the notation, that in this example the charge current flows in the x direction and has only an x component (compare Fig. 11.1 with Figs. 2.1 or 2.2).

For the spin current the situation is actually simpler. There we did not drop (see discussion below Eq. (9.9)) the zero voltage terms in the final expressions, because a spin current may exist at zero voltage. Of course these terms remain, while the positive voltage terms now get dropped, greatly simplifying the calculations. The spin currents are now driven by the phase difference, not by the voltage. They depend on the variable that we now call x, and this is the variable with respect to which spatial derivatives are taken. On the other hand we must remember that the spin current remains a vector *in spin space*. The components there refer to the axis of

spin quantization, which remains the z-axis. Similar considerations apply
to the local magnetization $\mathbf{m}(x)$, see Eq. (9.3).

The spin transfer torque $\boldsymbol{\tau}(x)$ can again be evaluated from Eq. (9.6). It
is related to the derivatives of the spin current via Eq. (9.5) which in the
steady state can be written, as in Eq. (9.10):

$$\tau_i = \frac{\partial S_i}{\partial x}. \tag{11.3}$$

Given the spin current, one could use Eq. (11.3) to evaluate the STT. It
is however a good safety measure to occasionally evaluate both sides of
Eq. (11.3) independently and use this equation as a consistency check. We
have seen examples of this checking e.g., in the insets of Fig. 9.2. However,
the direct use of Eq. (9.6), does have the advantage of avoiding the numerical derivatives that arise when using the right side of Eq. (11.3). Additional
physical insight can be gained by integrating Eq. (11.3) over a particular
region, e.g., F_1:

$$S_x(b) - S_x(a) = \int_{F_1} dx \tau_x = \tau_{x,\text{tot}}, \tag{11.4}$$

which means that the change in spin current through F_1 (from $x = a$ to
$x = b$) is equivalent to the net torque acting within those boundaries.

11.2.2   Triplet amplitudes and spin rotations

Of course, as in the simpler cases previously considered, triplet pairs
will form whenever the internal fields of the magnetic layers are non-
collinear. These triplet pairs will modify the current-phase relations. The
triplet amplitudes are still conveniently given formally by the expressions in
Eqs. 3.1 or 3.7.

As already mentioned in Chap. 6 it is sometimes necessary, or just
convenient, to evaluate the triplet amplitudes along a different spin axis.
For example, one may wish to use the direction of the local magnetization
in one of the F layers as the axis of quantization. To do so one rotates the
quantization axis so that it is aligned with the local magnetization direction
using the spin rotation matrices as seen in the discussion of Eq. (6.3). Here
we show how to perform the spin rotations for the two triplet components f_0
and f_1 (the singlet amplitude is of course invariant under spin rotations).
The problem simplifies if all one wishes is to align the spin quantization
axis with a local magnetization direction: this is sometimes desirable as
it affords easier physical interpretation of the results. The central quantity

that is to be used to perform the desired rotations is the spin transformation matrix $\mathcal{T}$ in particle-hole space.

The quasiparticle amplitudes transform as,

$$\Psi'_n(x) = \mathcal{T}\Psi_n(x). \tag{11.5}$$

The matrix $\mathcal{T}$ can be written as:

$$\mathcal{T} = \begin{bmatrix} \mathcal{A} & 0 \\ 0 & \mathcal{B} \end{bmatrix}, \tag{11.6}$$

where the submatrices $\mathcal{A}$ and $\mathcal{B}$ are trigonometric functions solely of the angles that describe the local magnetization orientation. Expressing the orientation of the exchange fields in the regions F_1 and F_2 in terms of the angles θ_i and ϕ_i introduced in Eq. (11.1) and Fig. 11.1 we can write $\mathcal{A}$ and $\mathcal{B}$ as the following 2×2 matrices:

$$\mathcal{A} = \begin{bmatrix} \cos\dfrac{\phi_i}{2}\sin\theta_i^+ + i\sin\dfrac{\phi_i}{2}\sin\theta_i^- & -\cos\dfrac{\phi_i}{2}\sin\theta_i^- - i\sin\dfrac{\phi_i}{2}\sin\theta_i^+ \\ \cos\dfrac{\phi_i}{2}\sin\theta_i^- - i\sin\dfrac{\phi_i}{2}\sin\theta_i^+ & \cos\dfrac{\phi_i}{2}\sin\theta_i^+ - i\sin\dfrac{\phi_i}{2}\sin\theta_i^- \end{bmatrix}, \tag{11.7}$$

$$\mathcal{B} = \begin{bmatrix} \cos\dfrac{\phi_i}{2}\sin\theta_i^+ - i\sin\dfrac{\phi_i}{2}\sin\theta_i^- & \cos\dfrac{\phi_i}{2}\sin\theta_i^- - i\sin\dfrac{\phi_i}{2}\sin\theta_i^+ \\ -\cos\dfrac{\phi_i}{2}\sin\theta_i^- - i\sin\dfrac{\phi_i}{2}\sin\theta_i^+ & \cos\dfrac{\phi_i}{2}\sin\theta_i^+ + i\sin\dfrac{\phi_i}{2}\sin\theta_i^- \end{bmatrix}, \tag{11.8}$$

where $\theta_i^\pm \equiv \theta_i/2\pm\pi/4$. Using the spin rotation matrix $\mathcal{T}$, one can transform the original BdG equations Eq. (2.8) by performing the unitary transformation: $\mathcal{H}' = \mathcal{T}\mathcal{H}\mathcal{T}^{-1}$ (of course we have $\mathcal{T}^\dagger\mathcal{T} = 1$). We then end up with the magnetization effectively along the new z-axis and:

$$\mathcal{H}' = \begin{pmatrix} \mathcal{H}_0 - h & 0 & 0 & \Delta \\ 0 & \mathcal{H}_0 + h & \Delta & 0 \\ 0 & \Delta^* & -\mathcal{H}_0 + h & 0 \\ \Delta^* & 0 & 0 & -\mathcal{H}_0 - h \end{pmatrix}. \tag{11.9}$$

A technical benefit of working in such a rotated coordinate system is that now the Hamiltonian matrix can be reduced to a smaller 2×2 size by using symmetry properties that now exist between the quasiparticle amplitudes and energies. As is the case under all unitary transformations, the eigenvalues here are preserved, but the eigenvectors are modified in general according to Eq. (11.5). Thus, for example, operating on the wave-functions using Eq. (11.5), and examining the terms involved in calculating the singlet pair correlations, it is rather easily shown that for a given set of quantum numbers n and position x, the following relation between the transformed (primed) and untransformed quantities holds: $u'_{n\uparrow}v'^{*}_{n\downarrow}+u'_{n\downarrow}v'^{*}_{n\uparrow}=u_{n\uparrow}v^{*}_{n\downarrow}+u_{n\downarrow}v^{*}_{n\uparrow}$. Thus, the terms that dictate the singlet pairing are invariant for any choice of quantization axis, transforming as scalars under spin rotations, as they should.

The terms governing the triplet amplitudes on the other hand are generally not invariant under spin rotations. It is illuminating to see how both the equal-spin and different spin triplet correlations transform. The relevant particle-hole products in Eq. (3.7) that determine f_0, upon the spin transformations obey the following relationships:

$$u'_{n\uparrow}v'^{*}_{n\downarrow} - u'_{n\downarrow}v'^{*}_{n\uparrow} = \cos\theta_i \left(u_{n\uparrow}v^{*}_{n\uparrow} + u_{n\downarrow}v^{*}_{n\downarrow}\right) \qquad (11.10)$$
$$+ \sin\theta_i \left[\cos\phi_i\left(u_{n\uparrow}v^{*}_{n\downarrow} - u_{n\downarrow}v^{*}_{n\uparrow}\right) + i\sin\phi_i\left(u_{n\uparrow}v^{*}_{n\uparrow} - u_{n\downarrow}v^{*}_{n\downarrow}\right)\right].$$

Similarly the quasiparticle terms in the sum for f_1 (in Eq. (3.7)) transform as:

$$u'_{n\uparrow}v'^{*}_{n\uparrow} + u'_{n\downarrow}v'^{*}_{n\downarrow} = \sin\theta_i \left(u_{n\uparrow}v^{*}_{n\uparrow} + u_{n\downarrow}v^{*}_{n\downarrow}\right) \qquad (11.11)$$
$$+ \cos\theta_i \left[\cos\phi_i\left(u_{n\downarrow}v^{*}_{n\uparrow} - u_{n\uparrow}v^{*}_{n\downarrow}\right) + i\sin\phi_i\left(u_{n\downarrow}v^{*}_{n\downarrow} - u_{n\uparrow}v^{*}_{n\uparrow}\right)\right].$$

Thus the triplet amplitudes f_0 and f_1 in the rotated system are linear combinations of the f_1 and f_0 in the original unprimed system (and vice versa). It is a simple matter to go from the rotated to the unrotated system (and vice versa) by the route expressed in Eq. (11.5) and the above discussion.

11.3 Results

11.3.1 *General*

Some results of a systematic investigation of the situations discussed above will be now presented. The figures are from Ref. [34] where results for additional cases can be seen. The results are given in terms of the same convenient dimensionless quantities as in previous chapters, see in particular Table 2.1. For the superconducting correlation length ξ_0 the results are for the value $k_F\xi_0 \equiv \Xi_0 = 100$, and the computational region occupied by each of the S electrodes corresponds to a width of $8\xi_0$ as mentioned above.

The low temperature regime only is considered, although it is not particularly difficult to extend the results to finite temperature: a certain number of rather obvious Fermi function factors must be added to the definitions of the currents and the magnetization, of course. For rather sound reasons, the strength of the magnetic exchange fields, h, is taken to be the same in all the magnets: its dimensionless value is set to $h = 0.1$. Results will be shown for a variety of orientation angles of the magnetic exchange field in each of the F regions, depending on the quantity being studied. Spin currents, magnetization and torque components etc., are normalized as in previous chapters, in particular see Chap. 9, especially the remarks at the end of Sec. 9.2. When presenting results for the currents, we normalize the charge current densities $j_x(x)$ by J_0, where $J_0 \equiv e n_e v_F$, and $v_F = k_F/m$ is the Fermi velocity. Interfacial scattering represented by delta functions of dimensionless strengths H_{Bi}, in the same context and notation as in Eq. (6.2), will be included at all the interfaces, of which there are up to four. The self-consistency of the pair potential is always taken into account: it ensures an accurate representation of the Cooper pair correlations throughout the system, as it is the ultimate source of the proximity effects. The pair potential depends to varying degrees on the parameters outlined above. In some cases the dependence is rather obvious: for instance, large H_B values result in weaker proximity effects. In other cases it is more intricate and will be analyzed more carefully.

11.3.2 *Current-phase relations*

I begin by showing some results for the self-consistent current phase relations in a simple double layer ferromagnet Josephson junction, an SF_1F_2S structure. At the interfaces between the F and S regions, quasiparticles undergo Andreev and conventional reflections. The superposition of these waves in the two F regions results in subgap bound states that contribute, together with the continuum states, to the total current flow.

Results for this structure are in Fig. 11.2. In the example shown, we see the current phase relation (CPR) for the case where the two ferromagnets have unequal widths: $D_{F1} = 10$ and $D_{F2} = 100$. This asymmetric choice of widths helps ensure that equal-spin triplet correlations are generated in the system. The angular parameters in this figure are fixed at $\theta_1 = 90°$, and $\theta_2 = 90°$, corresponding to in-plane magnetization orientations. The F_1 layer has its magnetization aligned in the z direction ($\phi_1 = 0°$). The first two panels, (a) and (b), display three different

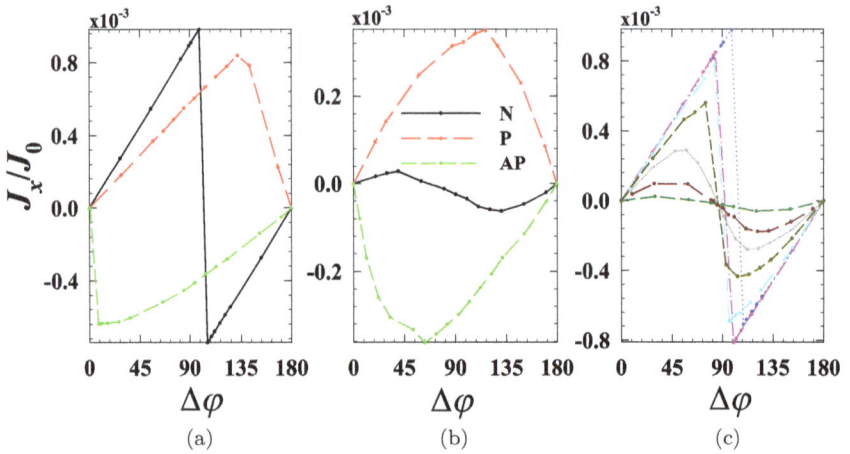

Fig. 11.2 Normalized (see text) Josephson current vs. phase difference, $\Delta\varphi$, for a SF_1F_2S structure with $D_{F1} = 10$, $D_{F2} = 100$, and $h = 0.1$. For panels (a) and (b), the legend in (b) labels the relative in-plane magnetization orientations: parallel (P), antiparallel (AP), or normal (N). Two interface scattering strengths are considered: (a) $H_B = 0$, and (b) $H_B = 1$. In panel (c), the magnetization orientations are fixed in the orthogonal configuration ($\phi_1 = 0$, $\phi_2 = 90°$, see Fig. 11.1), and the interface scattering is varied as $H_B = 0, 0.2, 0.5, 0.6, 0.7, 0.8, 1$ (in descending order of peak heights).

relative in-plane magnetization configurations in the F_2 layers: parallel (P) ($\phi_2 = 0°$, see Fig. 11.1), antiparallel (AP) ($\phi_2 = 180°$), and normal (N) ($\phi_2 = 90°$). A look at Fig. 11.1) will make this clearer. Please do not confuse the letter N, when it denotes the normal configuration, with its separate italicixed counterpart N denoting a "normal" layer. It is always obvious from the context. Two different strengths of the interface scattering parameter are considered. In (a) there is no interface scattering ($H_B = 0$), while in (b), a rather high rate of scattering is present, with $H_B = 1$ at all interfaces. The CPR for the collinear configurations (P or AP) possesses the conventional 2π periodicity, and the supercurrent flows oppositely for the two alignments. Recall that in these configurations there are no $m = \pm 1$ triplet pairs. When the relative in-plane magnetizations are orthogonal to each another, the CPR becomes π periodic as revealed by the sawtooth-like pattern in (a) or the more rounded behavior in (b), both of which change sign at $\Delta\varphi \approx 90°$. This is a consequence of the emergence of equal-spin triplet correlations, which are now allowed. When strong interface scattering is present, Fig. 11.2(b) shows that the π periodic CPR (N case) is substantially diminished, relative to the P or AP cases. This is because the proximity effect is weakened, with a resulting reduction of the associated equal-spin triplet correlations. Further elucidation of the effects that interface scattering on this π-periodic supercurrent, is provided in Fig. 11.2(c), which shows results for the same SF_1F_2S junction with varying degrees of scattering strengths H_B and with the relative in-plane magnetizations fixed and orthogonal to each other ($\phi_2 = 90°$). Increasing H_B clearly leads to a crossover in the CPR from a sawtooth to sinusoidal form and to a marked reduction of the supercurrent flow. As this occurs, the phase difference $\Delta\varphi$ yielding the critical current density also declines.

Next let us see an example of the case where one includes a nonmagnetic normal metal "spacer" separating the two ferromagnets. As we have seen in Chap. 6 such spacers are experimentally convenient, or even necessary, when it is wished [21] to rotate the magnetization in one magnet only. The example shown in Fig. 11.3 focuses on the case where triplets are allowed: the in-plane mutual magnetizations are kept orthogonal, (with $\phi_2 = 90°$, and $\theta_2 = 90°$) when the CPR curves become strongly non-sinusoidal.

In panels (a) and (b) of Fig. 11.3 the ferromagnet widths are $D_{F1} = 10$ and $D_{F3} = 100$, while the width of the normal metal spacer, D_N, varies

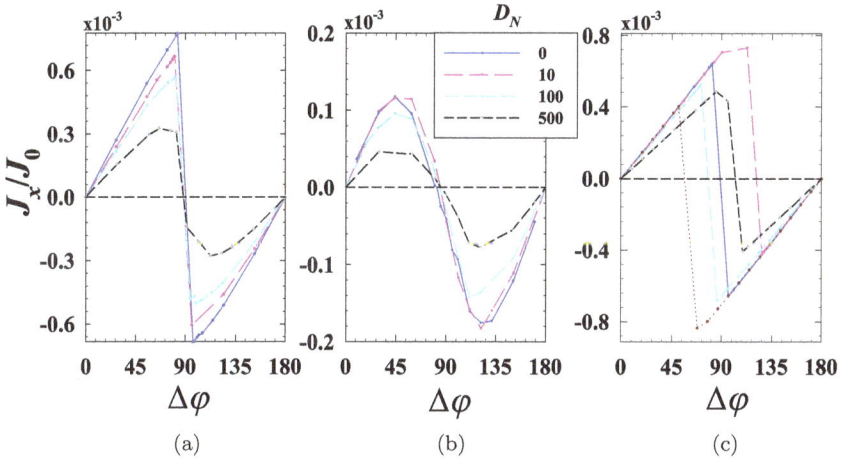

Fig. 11.3 Current-phase relationship for a SF_1NF_3S structure. The legend in panel (b) labels the N spacer widths, D_N. The peak heights decrease as D_N increases. The relative in-plane magnetization angle between the two F layers is 90°. For panels (a) and (b), the layers F_1 and F_3 have widths $D_{F1} = 10$, and $D_{F3} = 100$. In panel (a) $H_B = 0.5$, in panel (b) $H_B = 0.8$. In panel (c), $H_B = 0$, with ferromagnet widths $D_{F1} = 10$, and $D_{F3} = 380$. The additional dotted curve in this last panel corresponds to $D_N = 5$, illustrating the sensitivity of the CPR to D_N.

from zero to 500 (corresponding to $0 \leq D_N/\Xi_0 \leq 5$) as indicated in the
legend in panel (b). The interfacial scattering parameter is $H_B = 0.5$ for
panel (a) and $H_B = 0.8$ in (b), at all interfaces. In either case, increasing
D_N leads to a marked overall reduction in the supercurrent flow. Comparing panels (a) and (b) we see that for larger interface scattering, the CPR
becomes less sensitive to D_N. A wider junction, with ferromagnet widths
$D_{F1} = 10$ and $D_{F3} = 380$, is shown in panel (c), with $H_B = 0$. The details
of the sawtooth CPR reveal that the supercurrent flow can be quite sensitive
to the spacer width, as the value of $\Delta\varphi$ where the current abruptly reverses
direction varies considerably for incremental changes in D_N. In this panel,
results for an additional smaller value of the spacer thickness, $D_N = 5$, are
also shown. The π-periodic CPR is closely related to the generation of the
equal-spin triplet correlation, as we will see in more detail in Sec. 11.3.4
below. Also, one can infer from Fig. 11.3 that the introduction of an additional non-magnet metallic layer can quantitatively changes the transport
properties. Connections between the results in this figure and those for the
induced triplet pair correlations which will be shown in Figs. 11.7 and 11.8
will be made in Sec. 11.3.4.

Now, let us examine the effects of increased magnetic inhomogeneity. This we do in Fig. 11.4 where we can see results for a five layer

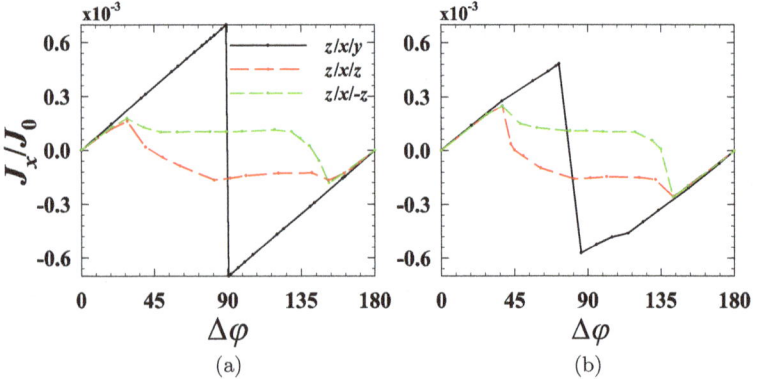

Fig. 11.4 Normalized Josephson current vs. phase $\Delta\varphi$ for a $SF_1F_2F_3S$ structure. The
legend labels the three relative magnetization directions of the F layers which are
included: as far as the layers F_1 and F_3 these correspond to the configurations labeled
N ($z/x/y$), P ($z/x/z$), and AP ($z/x/ - z$) in the previous figures, while the intermediate layer F_2 always has its magnetization in the out-of-plane (x) direction. The F_3
layer has in all cases width $D_{F3} = 380$. For the other F layers, widths are (panel (a))
$D_{F1} = D_{F2} = 10$, and (panel (b)) $D_{F1} = D_{F2} = 20$.

$SF_1F_2F_3S$ junction. This structure is complementary to both that considered in Fig. 11.3, with the normal layer changed to a magnetic one, and to that in Fig. 11.2, where the main difference is the additional ferromagnet layer. In the present case, this intermediate layer has a magnetic exchange field in the out-of-plane direction, the x-axis in this chapter. This corresponds to $\theta_2 = 0°$ in Fig. 11.1. The two outer layers, F_1 and F_3 are in parallel or antiparallel configurations. The relative magnetic orientations plotted are labeled in the legend (panel (b)) by the directions of the axes along with the magnetizations are aligned in each F layer. For example, $z/x/y$ denotes a sample in which F_1 and F_3 are normal to each other (the configuration labeled N in previous figures) contrasting with the out-of-plane magnetization in F_2. The width of the F_2 layer is identical to that of the F_1 layer: $D_{F1} = D_{F2} = 10$ in (a), and $D_{F1} = D_{F2} = 20$ in (b). The thicker ferromagnet F_3 has width $D_{F3} = 380$. The π-periodic CPR that arises in double magnet SF_1F_2S junctions with orthogonal in-plane magnetizations, as seen in Fig. 11.2, remains relatively unchanged by the additional out-of-plane intermediate ferromagnet. However, it was seen in Figs. 11.2(a)–(c) that the collinear P or AP magnetic states, while showing 2π periodicities, behave in a roughly piecewise linear fashion and that the current maintains its direction throughout the whole $\Delta\varphi$ range. Now, on the other hand, we see that the insertion of the additional F layer between the collinear ferromagnets, which makes it possible for equal-spin triplet correlations to be generated, leads as expected to a profoundly modified supercurrent. For phase differences in the vicinity of 0 or π, Figs. 11.4(a) and (b) show that the current is approximately linear in the phase difference, but there is a broad intermediate range of $\Delta\varphi$, where the supercurrent flow is relatively uniform. Moreover, for either the P or AP configuration, varying the phase can result in the Josephson current switching direction. These trends are the same for each of the cases shown in (a) or (b).

11.3.3 *Magnetic orientation and CPR*

In the above subsection, Sec. 11.3.2, we have vividly seen that because of triplet pairs generation, magnetization orientation has a strong effect on the current phase relations. We have seen this for a few different ferromagnetic Josephson junction configurations, Clearly, it now is appropriate to study in more detail the the changes produced in the CPR by varying magnetization orientations. To do this, it is convenient to set the macroscopic phase difference to a prescribed value and study the supercurrent response within

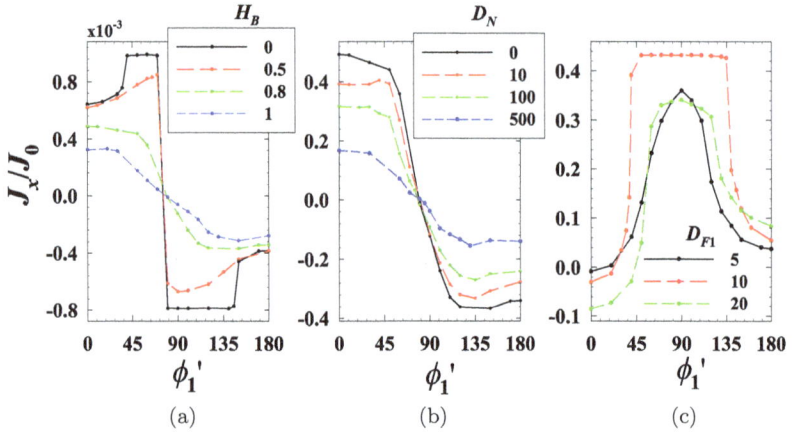

Fig. 11.5 Normalized Josephson current vs. the in-plane relative magnetization angle $\phi_1' \equiv \phi_1 + 90°$ (see text and Fig. 11.1). In panel (a) results are for a SF_1F_2S junction, with $\Delta\varphi = 100°$, $D_{F1} = 10$, $D_{F2} = 100$, and interfacial scattering as in the legend (current size decreases with increasing H_B). In panel (b), they are for an SF_1NF_3S junction with $\Delta\varphi = 100°$, $D_{F1} = 10$, $D_{F3} = 100$, D_N as in the legend (current decreases with increasing D_N), and $H_B = 0.8$. In panel (c) a simple SF_1F_2S junction is considered with $H_B = 0$, $\Delta\varphi = 60°$, $D_{F2} = 380$, and three different D_{F1} values (see legend). The $\mathbf{h}$ field in F_2 or F_3 is always along y.

the junctions for a range of relative magnetization orientations. This is how we will show results now.

Let us return first to the basic SF_1F_2S Josephson junction of Fig. 11.2 but now with a phase difference set at $\Delta\varphi = 100°$. In Fig. 11.5(a), we can see plots of the normalized supercurrent density as a function of the in-plane ($\theta_1 = 90°$) magnetization angle ϕ_1', where $\phi_1' \equiv \phi_1 + 90°$, with ϕ_1 being the angle shown in Fig. 11.1. The magnetization in F_2 is set along the y direction, therefore the P and AP states correspond to $\phi_1' = 0°$, and $\phi_1' = 180°$ respectively. Four interface scattering strengths are considered in panel (a), as indicated in its legend. In all cases, by tuning the relative alignment angles, supercurrent switching occurs when the mutual magnetization orientations are approximately orthogonal ($\phi_1 \approx 90°$). As expected, the supercurrent flow is greatest for transparent interfaces ($H_B = 0$), and decreases with increasing H_B. As this decrease occurs the sensitivity to ϕ_1' becomes weaker. The maximum current flow occurs at different ϕ_1' values, depending on the interface scattering strength. In Fig. 11.5(b), an additional N layer, of variable width as indicated, is inserted between the two ferromagnets and the interface scattering parameter is set to $H_B = 0.8$, at the same $\Delta\varphi = 100°$ value as in panel(a). The solid $D_N = 0$ curve is the same as the $H_B = 0.8$ curve in panel (a) (third from the top at small ϕ_1'). This panel shows that increasing the N layer thickness tends to generally dampen the current through the junction. The current flow peaks when the two ferromagnets are aligned (P configuration); it vanishes altogether near $\phi_1 = 90°$ and then reverses direction as the the AP state is approached. To complete this illustration, in Fig. 11.5(c) the supercurrent flow is plotted for a thick SF_1F_2S junction, with $D_{F2} = 500$, and three F_1 widths (see legend). The phase difference is now set at $\Delta\varphi = 60°$. For this geometry, we see that the current has some features that contrast with the previous cases involving thinner F layers. Now, for instance, when tuning ϕ_1', the weakest current flow occurs in the P configuration, and it is low also in the AP case, while the maximum occurs in the orthogonal configuration. For all F widths considered, the current undergoes rapid changes in the vicinity of the middle value between the N and P or the N and AP configurations, $\phi_1 = 45°$ and $\phi_1 = 135°$, with the $D_{F1} = 10$ case changing the most, and then remains nearly constant at angles near the P and AP orientations. Since $\Delta\phi = 60°$ is near where the maximum currents occur in π-periodic Josephson junctions, it is evident that the results in Fig. 11.5(c) are *a direct consequence of the induced equal-spin triplet correlations.* We will see below in Sec. 11.3.4 more about how these results correlate with the triplet generation behavior.

Let us study next the magnetization orientation role in the supercurrent for more complicated $SF_1F_2F_3S$ structures. To do this, consider a scenario where the F_1 and F_3 layers have magnetizations orthogonal to each other, along the z and y directions respectively, while the magnetization in the central F layer rotates on the $x-z$ plane ($\phi_2 = 0$) from along the x-axis (which, we recall, is normal to the layers in this chapter) to the z-axis. Figure 11.1 should be kept in mind. In Fig. 11.6(a), the normalized current density is plotted as a function of θ_2 for several F_3 layer widths, D_{F3}. The thicknesses of the outer F layers are fixed at $D_{F1} = D_{F2} = 10$, and values of D_{F3} are chosen so that they lead to both symmetric and asymmetric plots and structures. In each case, $\Delta\varphi = 100°$, and interface scattering is present with $H_B = 0.8$ at all interfaces, except for the widest junction with $D_{F3} = 380$, where $H_B = 0$. In the case $D_{F1} = D_{F2} = D_{F3} = 10$, where all F thicknesses are identical, the current flow is approximately antisymmetric around $\theta_2 = 180°$. The current flow is suppressed for the asymmetric $D_{F3} = 100$ situation by decoherence effects arising from the larger width. While j_x for the SF_1F_2S junctions in Fig. 11.5 was π periodic

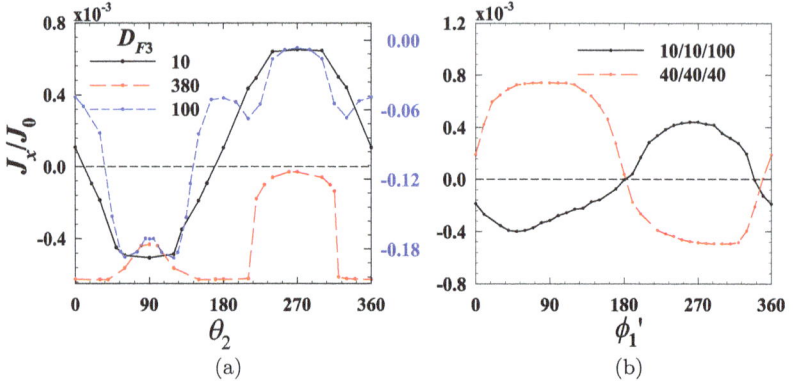

Fig. 11.6 Normalized Josephson current for a $SF_1F_2F_3S$ junction. Panel (a): vs. the out-of-plane angle θ_2 (see Fig. 11.1). Panel (b): vs. the in-plane angle $\phi_1' = \phi_1 + 90°$. The phase difference is $\Delta\varphi = 100°$ for both cases. In (a) the ferromagnets, F_1 and F_2 have widths $D_{F1} = D_{F2} = 10$, and the legend lists the three F_3 widths considered. The exchange field points along the y direction in F_3 and the z direction in F_1. The right vertical scale is for the $D_{F3} = 100$ (dashed blue curve) case only. For $D_{F3} = 380$, there is no interface scattering ($H_B = 0$), while for the remaining D_{F3} cases $H_B = 0.8$. In (b) $\theta_1 = 90°$, and **h** is along y in F_2, and along z in F_3. The legend indicates the F layer widths of the two structures considered: they have the same total width. Interfacial scattering strength in both cases is $H_B = 0.8$.

in ϕ_1', variations in θ_2 for trilayer ferromagnetic junctions are in general 2π periodic, as seen in this figure. The asymmetric case also exhibits a more intricate plot structure as the magnetization angle is swept. For the narrower junction, $D_{F3} = 10$, the orientations in which the current switches direction are near $\theta_2 = 0°$ and $\theta_2 = 180°$. When $D_{F3} = 100$, the current never changes direction (see the right side vertical scale). The rotating out-of-plane exchange field of F_2 does affect the strength of the current with 2π periodicity. For the highly asymmetric $D_{F3} = 380$ case, the current again maintains its flow direction over the full range of θ_2. It is also approximately constant except for orientations when the exchange field in F_2 points near the z-axis: $\theta_2 = 90°$ or $\theta_2 = 270°$. For orientations near $\theta_2 = 270°$, the current is strongly suppressed.

To display another aspect of the same phenomenon, in Fig. 11.6(b) the orientation of the in-plane magnetization in the first ferromagnetic layer is now varied, while the magnetization directions in the other two F layers are kept fixed: for F_2 along y, and for F_3 along z, an N (orthogonal) configuration. Results for two structures are shown in this panel, with the total width of the three ferromagnetic regions being kept constant: in one case, $D_{F1} = D_{F2} = 10$, and $D_{F3} = 100$, while in the other one, $D_{F1} = D_{F2} = D_{F3} = 40$. Since all three F layers have in-plane magnetizations, the current has now a predominant period of 2π in ϕ'_1. Both cases exhibit similar behavior as a function of ϕ'_1. The current vanishes when the internal field in F_1 is antiparallel to that in F_2 ($\phi_1 = 180°$), and is highest when the field in F_1 lies nearly in between those of F_2 and F_3. Thus, the charge supercurrent which flows oppositely in the two structures, can be effectively switched on or off by manipulating the in-plane magnetization angle of the first ferromagnet.

11.3.4 *Induced triplet pairing in ferromagnetic Josephson junctions*

It is now time to discuss, as previously promised several times, the effects induced triplet pairing correlations in these ferromagnetic Josephson junctions. The presence of multiple misaligned ferromagnets yields, as we have seen in Chap. 3, particularly Secs. 3.4 and 3.5, both $m = 0$ and the $m = \pm 1$ triplet pair amplitudes, defined in Eq. (3.7), as permitted by conservation laws and the Pauli principle.

To gain an overall view of the opposite-spin triplet amplitudes, f_0, and the equal-spin amplitudes, f_1, we can look at Fig. 11.7, where there are plots of the spatial behavior (the X dependence) of these correlations in the N and F regions of an SF_1NF_3S junction. The plots are of the real parts of these complex quantities, as given in Eq. (3.7) but one can keep in mind that the imaginary components obey similar trends. By fully taking into account the microscopic proximity effects between all of the interacting layers, we can learn that the interference effects taking place over the Fermi length scale combined with the self-consistency of the pair potential, create the interesting local spatial behavior.

This figure requires some explanation for its understanding. The range of X included comprises the F_1, N, and F_3 regions. Recall that $D_S = 800$. The range of X is divided into two subranges, one for F_1 and the other for N and all or most of F_3 combined. The geometrical parameters in this figure are $D_{F1} = 10$, $D_{F3} = 380$, as used in Fig. 11.3(c). The value of the spacer thickness is $D_N = 10$ (top panels) or $D_N = 100$ (bottom panels). Thus, in all panels the region $800 < X < 810$ is occupied by F_1 while F_3 occupies the region $820 < X < 1200$ in the top panels (a) and (b) and $910 < X < 1290$ in the bottom panels (c) and (d), where the vertical dotted line identifies the N/F_3 boundary. Hence, different horizontal scales have

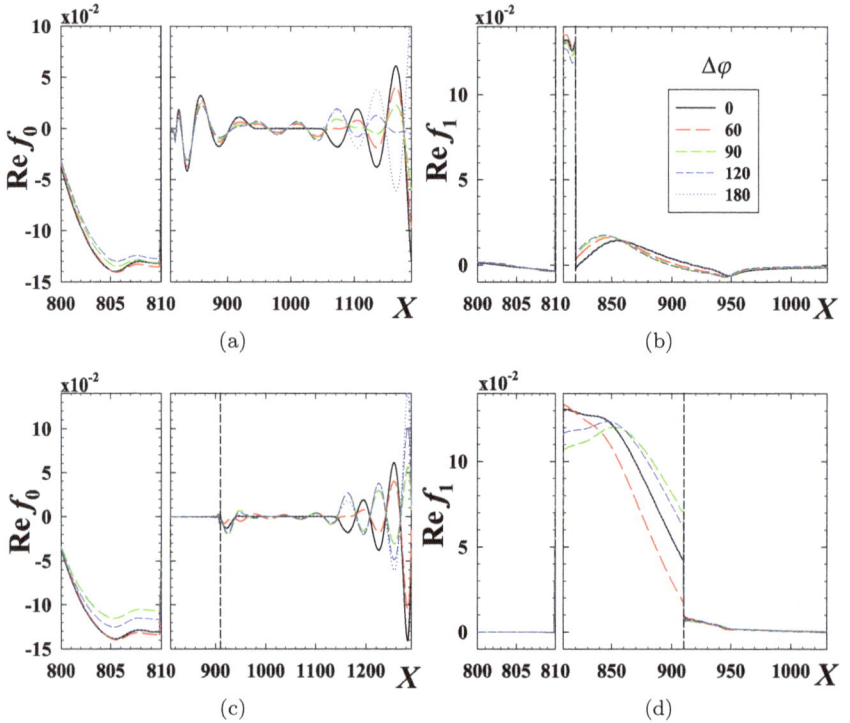

Fig. 11.7 Triplet correlations in an SF_1NF_3S Josephson junction as a function of position. The real parts of f_0 and f_1 (Eq. (3.7)) in the F and N regions are shown. The F layers have widths $D_{F1} = 10$ and $D_{F3} = 380$, as in Fig. 11.3(c). The same barrier values are also used. Note the different horizontal scale in the left plots (F_1 region) and the right plots (N and F_3 regions) and recall that the S layers (not shown) have $D_S = 800$. The top set of panels are for a normal metal spacer width $D_N = 10$, while the bottom set is for a thicker N layer, $D_N = 100$. The dashed vertical lines in panels (c) and (d) help visualize the interface between the N and F_3 regions. Only a portion of the F_3 region is shown in panel (d). Various phase differences $\Delta\varphi$ are considered (see legend in panel(b)). The magnetization in F_1 is along z, while it is along y in F_3.

to be used in each case. The scattering parameter is set to $H_B = 0$, and each curve corresponds to a different phase difference, $\Delta\varphi$ as indicated by the legend in panel (b). The exchange fields are in-plane and normal to each other: in F_1 $\mathbf{h}$ is along z, while in F_3 it is along y, thus we have a normal configuration which should come close to maximizing the triplets.

Examining the plots, we see that within the F_1 region, panels (a) and (c) show that the magnitudes of the f_0 pair correlations are approximately the same for both values of D_N, decreasing in the vicinity of the S/F interface located at $X = 800$, located at the left edge of the plots. The system with the wider normal metal layer, $D_N = 100$, is slightly more sensitive to phase variations. Within the ferromagnet F_3, the same panels (a) (c) show the oscillatory nature of f_0, which behaves similarly to the singlet pair amplitude, the periodicity arising from the difference in spin-up and spin-down wavevectors. For the chosen exchange field, the oscillations are limited in F_1 due to the confined width.

Turning now to the equal-spin triplet correlations f_1, panels (b) and (d) display behavior which contrasts with the f_0 results. In particular, within the narrow F_1 region, the f_1 triplet amplitudes are very small: the f_0 correlations clearly dominate. In the N region on the other hand, the f_0 correlations are very small, while the equal-spin triplet amplitudes are much larger, with a peak near the F_1/N interface (at $X = 810$), before dropping within the normal metal (see Fig. 11.7(c) and (d)). Finally, within F_3, the triplet amplitudes f_1 assume a slow, long-range variation contrasting with the damped oscillatory behavior of the f_0 curves. Thus we can conclude that in the nonmagnetic normal metal there is a large enhancement of triplet correlations consisting of equal-spin pairs regardless of whether the width of N is relatively thin or thick. For the thick N case, we can see in Fig. 11.3(c) that the maximum in the CPR occurs at $\Delta\varphi \approx 90°$. Remarkably, the corresponding f_0 is weakest in F_1 but f_1 is very uniformly distributed in N.

It is very instructive to also examine the spatially averaged triplet correlations as functions of the relevant system parameters. We have already used with profit such spatial averages in Chap. 6 and even more in Chap. 9; see Eq. (9.12) for a careful definition of these spatial averages. For example, in the top panels of Fig. 11.8 by tuning the relative angle of the in-plane internal fields, (e.g., varying ϕ'_1 at fixed $\phi_2 = 90°$), important overall features are revealed. The first two panels, (a) and (b), of Fig. 11.8 show the ϕ'_1-dependence of the magnitudes of the triplet amplitudes averaged over the F_3 region for the SF_1NF_3S structure and parameter values for which results were shown in Fig. 11.5(b). The same four representative D_N values are included as indicated in the legend. The proximity effects and hence the coupling between the two ferromagnets diminish with increasing D_N, and therefore the pair correlations become less sensitive to variations in ϕ'_1, as particularly observed for the largest $D_N = 500$ values. Other than the diminished magnitudes, the overall trends and behavior, however, do not depend strongly on the presence of the normal metal spacer. This may be important in experiment design, where spacers are usually needed.

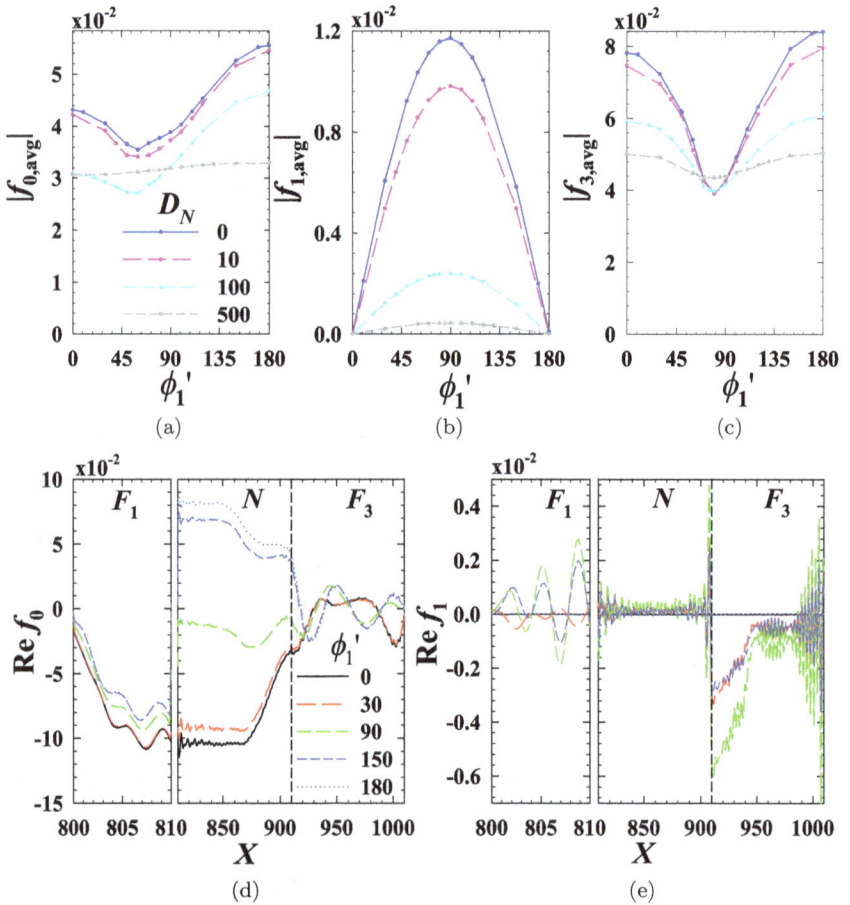

Fig. 11.8 Pairing amplitudes for a SF_1NF_3S structure. Top row: panels (a)–(b) have the normalized absolute values of the triplet amplitudes vs. in-plane relative magnetization angle $\phi'_1 \equiv \phi_1 + 90°$ averaged over the F_3 region. Panel (c) has, for comparison, the corresponding average for the singlet amplitude, which we denote as f_3 when it is complex. The geometrical and material parameters are the same as in Fig. 11.5(b), including the choices of D_N in the legend. The correlations decrease as D_N increases. The bottom panels (d) and (e) show the local spatial behavior of the triplet correlations for the $D_N = 100$ case studied in the top panels. The results for the two F regions are plotted in separate frames to make it easier to discern the triplet correlations in the narrow F_1 region. The dashed vertical lines mark the interface between the N and F_3 regions. Several values of ϕ'_1 are considered, as shown in the legend in (d). The internal field in F_3 is along y.

In panel (c) there is, for comparison, a plot of the absolute value of the ordinary singlet amplitude. This quantity we denote as f_3 ($f_3 \equiv \Delta/g$) in this chapter, as contrasted for example with the notation F in Eqs. 2.4 or 2.22, to emphasize that we are dealing now with a complex quantity. In performing this comparison, we see that the opposite-spin triplet correlations, f_0 and the singlet pair amplitude in (a) and (c) behave in rather similar ways, although f_3 is of course larger and also more symmetric about the orthogonal direction $\phi_1' = 90°$. Both contrast with f_1. When the relative magnetization orientation varies in inhomogeneous S/F systems, the process of singlet to triplet conversion plays a role in the transport and thermodynamic properties of such systems, as we have seen. It is apparent from panels (a) and (b) that the orientation ϕ_1' that leads to a minimum in f_0 (and also of f_3), corresponds to that where f_1 is largest. These occurrences arise when $\phi_1' \approx 90°$, that is when the exchange fields in F_1 and F_3 are nearly orthogonal. The value of this angle slightly shifts with variations in D_N, particularly for f_0. Changing D_N is seen to cause the optimal angle for local minima or maxima to differ from $\pi/2$. The exact angle for this can only be determined from a numerical self-consistent calculation as described here.

In the bottom panels, (d) and (e), of this Fig. 11.8 we can see the spatial dependence of the real parts of the triplet correlations throughout each of the three junction regions discussed in the top panels. Results for five different relative magnetic orientations are presented (see legend) including the P ($\phi_1' = 0°$), ($\phi_1' = 180°$), and N ($\phi_1' = 90°$) configurations. Consider carefully the regions plotted and the different scales. Examine first Fig. 11.8(d) in the F_1 region, the leftmost plot: we see there behavior similar to that found for the wider junction case in Fig. 11.7, including a relatively weak dependence on the orientation angle ϕ_1'. Recall that, in contrast, the dependence on the phase difference $\Delta\varphi$, as seen in Fig. 11.3 is quite strong. The H region, central plot, is most affected by variations in ϕ_1': the real part of f_0 changes sign when ϕ_1' is swept from the P to AP state, and nearly vanishes altogether at $\phi_1' \approx 90°$. In F_3, a series of oscillations are seen with a periodicity similar to that found in Fig. 11.7: this makes sense since an exchange field of the same strength is used in both figures, as throughout this chapter. In Fig. 11.8(e), the equal-spin f_1 amplitudes can be seen to exhibit a considerably different behavior. First, only three of the considered ϕ_1' values yield nonzero results, since the P and AP configurations cannot (see Chap. 3) generate equal-spin triplet correlations. Interestingly, within the F_1 region f_1 does not exhibit a slow decay, but rather oscillates with a period that is shorter than the oscillation period of opposite spin pairs, which is governed by the difference between spin-up and spin-down wave vectors. Within the normal metal layer, there is nearly a complete absence of equal-spin correlations — this is accompanied by the appearance of opposite-spin correlations, f_0 (see panel (d)). The f_1 amplitudes are largest in the F_3 layer, for the relative orientation of $\phi'_1 = 90°$, in agreement with the averaged results in Fig. 11.8(b). For magnetization orientation angles of $\phi'_1 = 30°$, and $\phi'_1 = 150°$, the f_1 amplitudes are identical due to the symmetry about $\phi'_1 = 90°$.

One can now correlate the features in Figs. 11.7 and 11.8 to those in Figs. 11.3 and 11.5. From Fig. 11.3, we learned that the N magnetic configuration often leads to the appearance of π-periodic Josephson junctions. From the behavior of the charge current in Fig. 11.5(b), we see that j_x vanishes at the value of ϕ'_1 where Fig. 11.8 shows that the averaged opposite-spin singlet and triplet correlations are smallest and f_1 is largest. Putting all of this together we are able to see concrete proof that in the N cases the average magnitude of the equal-spin triplet correlations is maximized. They are also weakly dependent on $\Delta\varphi$. Therefore, the CPRs for different magnetic configurations are essentially correlated with the detailed structure of their triplet correlations.

Let us consider next the behavior of the averaged triplet (and again for comparison singlet) amplitudes, as the magnetic orientation angles change. In Fig. 11.9, we examine these quantities as θ_2 changes (top panels), or ϕ'_1 change (bottom panels). This is done, as in Fig. 11.6 for the most complicated structure considered in this chapter, $SF_1F_2F_3S$ Josephson junctions. This discussion is therefore complementary to that done in connection with Fig. 11.6 involving the charge supercurrent. The geometric parameters are $D_{F1} = D_{F2} = 10$, and $D_{F3} = 100$. The region in which the pair correlations are averaged over is specified in the legends placed on top of the plots. In the top row of Fig. 11.9 are results corresponding magnetization orientations where θ_2 is swept over the entire angular range from $0°$ to $360°$, while the magnetizations are aligned along z in F_1 and along y in F_3. Therefore when $\theta_2 = 0°$ or $\theta_2 = 180°$, all three ferromagnets have mutually orthogonal magnetizations, corresponding to a high degree of magnetic inhomogeneity most favorable to triplet generation: one can expect that the equal-spin triplet correlations f_1 should be, on the average, at their highest values, while the opposite spin correlations should be weakest. Indeed, in the central region F_2, the opposite-spin singlet, f_3, and triplet f_0 correlations possess minima near these angles, in contrast to the spatially

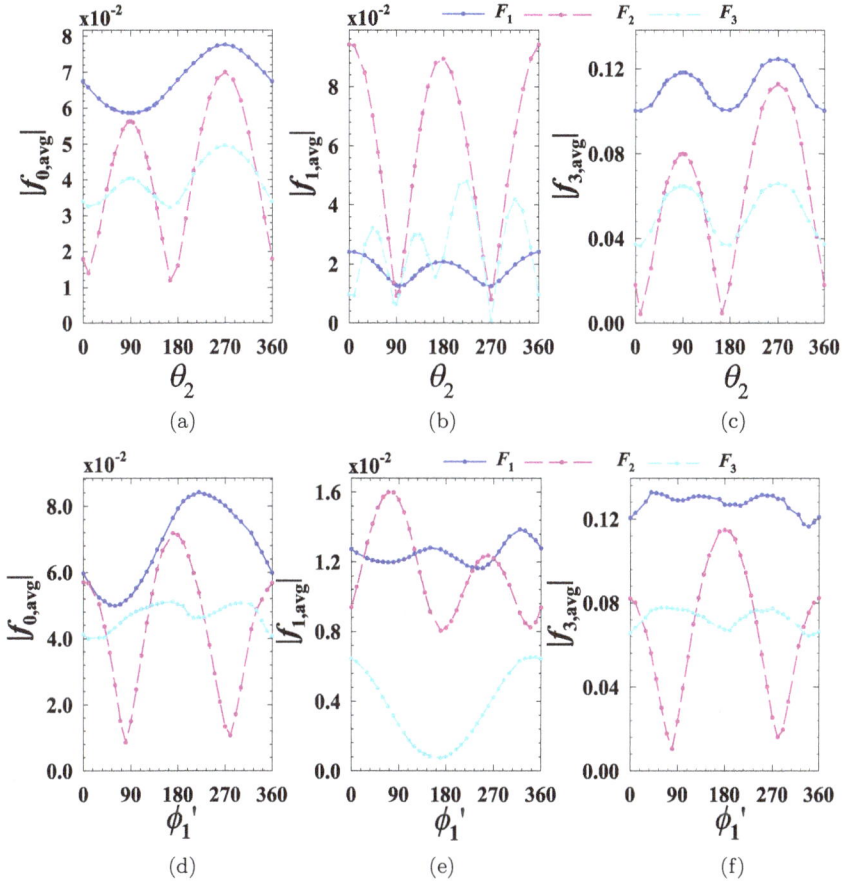

Fig. 11.9 Normalized triplet $|f_{0,\text{avg}}|$, $|f_{1,\text{avg}}|$, and singlet $|f_{3,\text{avg}}|$ amplitudes, averaged over the F regions indicated on the overhead legends, for $SF_1F_2F_3S$ Josephson junctions. These quantities are plotted as functions of θ_2 (top row) and (bottom row) ϕ_1' (recall Fig. 11.1). In the top row $\mathbf{h}$ is along z in F_1 and along y in F_3, while in the bottom row it is along y in F_2 and z in F_3. The geometrical and material parameters correspond to the $D_{F3} = 100$ cases presented in Fig. 11.6.

averaged f_1 amplitudes, which peak at those orientations. Although the general trends are usually the same for all F layers, geometrical effects can result in self-consistent triplet correlations with a more intricate and non-trivial structure, and this is the case with the averages over the F_1 and F_3 regions where the proximity effects are most prominent, since these layers are adjacent to S material. In the bottom panels of Fig. 11.9, the variation of the averaged triplet correlations with in-plane magnetization rotation of the F_1 layer is displayed. The other ferromagnets F_2 and F_3 have their magnetizations fixed in the y and z in-plane directions respectively. For this situation, the triplet opposite-spin f_0 amplitudes in F_1 are seen to be 2π periodic, peaking at $\phi_1' \approx 225°$. As in the previous row, there are simple correlations between the maximum and minimum values of f_0 and f_1 in F_2. These triplet amplitudes are seen to be largest when the relative orientations between F_1 and F_2 are either P ($\phi_1' = 0°, 360°$) or AP ($\phi_1' = 180°$). This is in agreement with the behavior of the triplet amplitudes found in two-magnet spin valve Chap. 8 systems [29].

11.3.5   *Spin transport*

Having now seen the salient features of supercurrent charge transport, and its connection with pair correlations in a variety of ferromagnetic Josephson junctions, we should next explore the spin degree of freedom and determine the crucial spin currents and the associated STT. Just as is we have seen it to be the case for Chap. 9 voltage driven current, the charge current driven by the macroscopic phase differences between the S electrodes drags spin with it, and can become spin-polarized when entering one of the ferromagnetic regions. This resulting spin current can then interact with the fields in the other ferromagnets and be modified by the local magnetizations due to the spin-exchange interaction, via the existence of STT. The conservation law associated with precess is described by Eqs. (9.6) and (11.3). It is important not only to understand the behavior of the spin-polarized currents in ferromagnetic Josephson junctions, but also the various ways in which to manipulate them for practical spintronic applications. With the insertion of N layers, the anisotropy energies between the magnetization of one of the F layers, typically pinned by an antiferromagnet, and that of a "free" F layer can be overcome via e.g., a small external field [21] as seen in Chap. 6. The STT can be studied [35] through the effect of the Josephson induced exchange interaction provided that the interaction energy exceeds the anisotropy energy. I therefore show results here for the equilibrium phase-driven spin currents and associated torques throughout the entire junction regions as functions of position, phase difference, and magnetization orientation angles.

The layered geometry implies, as repeatedly emphasized above, that the generally tensorial spin current is reduced to a vector in spin space, representing a spin vector current flowing in the spatial direction normal to the layers (the x direction in this chapter) and having in general three components in spin space. The normalization has been already explained. Because the internal field (and hence the torque) vanishes in the N and S regions, only the F regions of the junction can have a spatially varying spin current: in the N and S regions the spin current must be spatially invariant. Under the constant-phase but zero voltage boundary conditions the outer s-wave superconducting regions cannot support (see Chap. 9) a spin current, and hence $\mathbf{S}$ vanishes there. In a SF_1NF_3 junction, however, the intervening nonmagnetic N layer will couple the two ferromagnets via a constant spin current, which is related to the net torque acting within the F regions (see Eq. (11.4)). The spin current oscillates in the F_3 region. The amplitude of these oscillations depends on $\Delta\varphi$, while the period does not: the points in F_3 where S_x vanish are independent of $\Delta\varphi$. If a ferromagnet is very thin, as is our usual assumption for F_1, the spin currents vary nearly linearly with X; this variation can be viewed as a small segment of a sinusoidal function. To present an overall view of how the change in spin current and its associated torque vary as the phase varies it will be helpful to use the total torque, defined as the integral of the local torque, normalized as discussed above, over a layer. As it follows from Eq. (11.4) this integral is basically the total change in the spin current.

Specifically, let us now see in Fig. 11.10 how the spin currents and associated torques change when varying the relative exchange field directions between F_1 and F_3 in an SF_1NF_3S junction. A supercurrent is generated in the structure, at zero voltage, by maintaining a phase difference which is taken to be $\Delta\varphi = 100°$ in this figure. The orientation of the magnetization in F_1 is rotated in the $y-z$ plane, while that in F_3 is fixed along the y direction (as in Fig. 11.5(b)). Control of the free-layer magnetization can be achieved experimentally via external magnetic fields (Chap. 6), or spin-torque switching. In Fig. 11.10(a), the x component of the local spin current, $S_x(X)$, is shown throughout the junction's magnetic and normal layers as a function of position X, for four values of the ϕ'_1 angle (with $\theta_1 = \theta_3 = 90°$). In this phase-driven case, and under the orientation conditions plotted, the torque is predominantly in the x direction, and the variations of S_x are more prominent than those of the other components. This is why they are emphasized here. The spin current is seen to be a non-conserved quantity within the ferromagnets. This is as it should be, reflecting the existence of a STT. In the nonmagnetic normal metal connecting the two F regions, the current is constant and, as ϕ'_1 is varied, S_x cycles from positive to negative. To explore this further, let us examine the total change in spin current across F_1, as ϕ'_1 sweeps from the P to AP state.

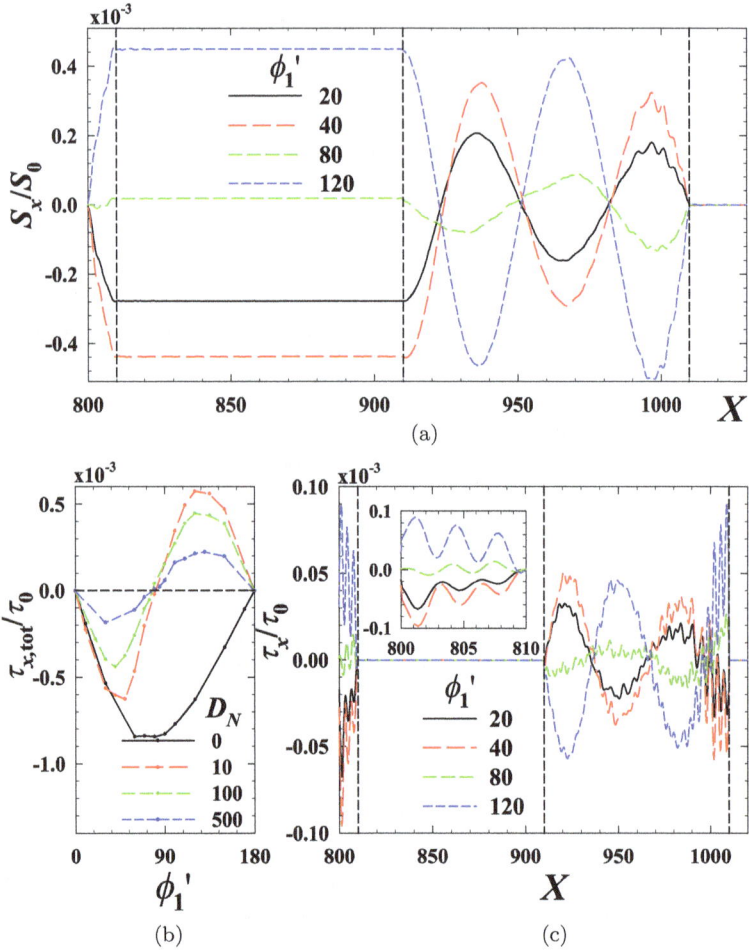

Fig. 11.10 Panel (a): Normalized x component of the spin current in an SF_1NF_3S Josephson junction as a function of position. The in-plane internal fields are orthogonal. The layer thicknesses are $D_{F1} = 10$, $D_N = 100$ and $D_{F3} = 100$. The interface scattering strength is set to $H_B = 0.8$. These values are the same as in Fig. 11.5(b). The dashed vertical lines mark the F_1/N and N/F_3 interfaces. The S regions are not included. Results for several values of ϕ'_1 (legend) are shown. Panel (b): the total torque $\tau_{x,\text{tot}}$ within the F_1 region as a function of the relative in-plane magnetization angle, ϕ'_1. Several normal metal widths are considered (see legend). The $D_N = 0$ results have been divided by 5 to fit the vertical scale. Panel (c): the x-component of the torque shown as a function of position for the same case considered in panel (a). An inset emphasizes the F_1 region.

This total change is related via Eq. (11.4) to the integrated torque in this region. Hence, in Fig. 11.10(b) there is a plot of $\tau_{x,\text{tot}}$ vs. ϕ'_1 for a wide range of D_N. When the normal metal spacer is absent ($D_N = 0$), the magnitude of the total torque reaches its peak around $\phi'_1 = 90°$, indicating that this component of the torque, which tends to align the magnetic moments of the two F layers is largest when they are mutually orthogonal. This makes eminent sense physically. The presence of even a thin normal metal spacer causes $\tau_{x,\text{tot}}$ to become much smaller (the $D_N = 0$ results are plotted after dividing them by five or they would not fit with the vertical scale) and nearly π-symmetric, so that now the orthogonal magnetic configuration produces negligible net torque within the F layers. Increasing D_N reduces the ferromagnetic coupling and hence reduces the magnitude of the mutual torques, although the π-periodicity is retained. Finally in Fig. 11.10(c) we can see a plot of the local value of the torque, and find that its spatial behavior is consistent with that of S_x as given in Eq. (11.3). The inset emphasizes the oscillations in the F_1 layer which would be invisible in the main scale. Comparing panels (a) and (c) of Fig. 11.10 we see that in F_3 current and torque oscillate in the same way, but out of phase, in keeping with Eq. (11.3).

To conclude this subsection, let us in Fig. 11.11 consider spin transport in the $SF_1F_2F_3S$ structure that we have looked at previously in Fig. 11.6(a). It has $D_{F1} = D_{F2} = 10$, and $D_{F3} = 100$. A phase difference of $\Delta\varphi = 100°$ maintains a charge current throughout the junction, and there is considerable interface scattering, with $H_B = 0.8$. The magnetization in F_1 is along z, and in F_3, it is along y, so they are in-plane and orthogonal to each other, with abundant triplets generated. The central ferromagnet, F_2, has a magnetization vector that has an out of plane component: it is rotated in the $x-z$ plane (see Fig. 11.1), so that for $\theta = 0°$, it is oriented along x, normal to the layers, and for $\theta = 90°$, it is aligned along z. For these more complex magnetic configurations, where one of the F layers possesses an out-of-plane exchange field, all three spin components of the current $\mathbf{S}$ or torque $\boldsymbol{\tau}$ should be considered. The top panels in Fig. 11.11 depict the components of the total torque, $\tau_{i,\text{tot}}$ ($i = x, y, z$) for each ferromagnet region, as identified in the legend above these panels. Since the total torque in a given direction equals (see Eq. (11.4)) the overall change in spin current, and, as explained above, there is no spin current in the S regions at fixed phase and no voltage, the sum of each component $\tau_{i,\text{tot}}$ over all F regions must be zero. This is seen in these three panels, where the oscillatory curves exactly cancel one another. For each of the three components,

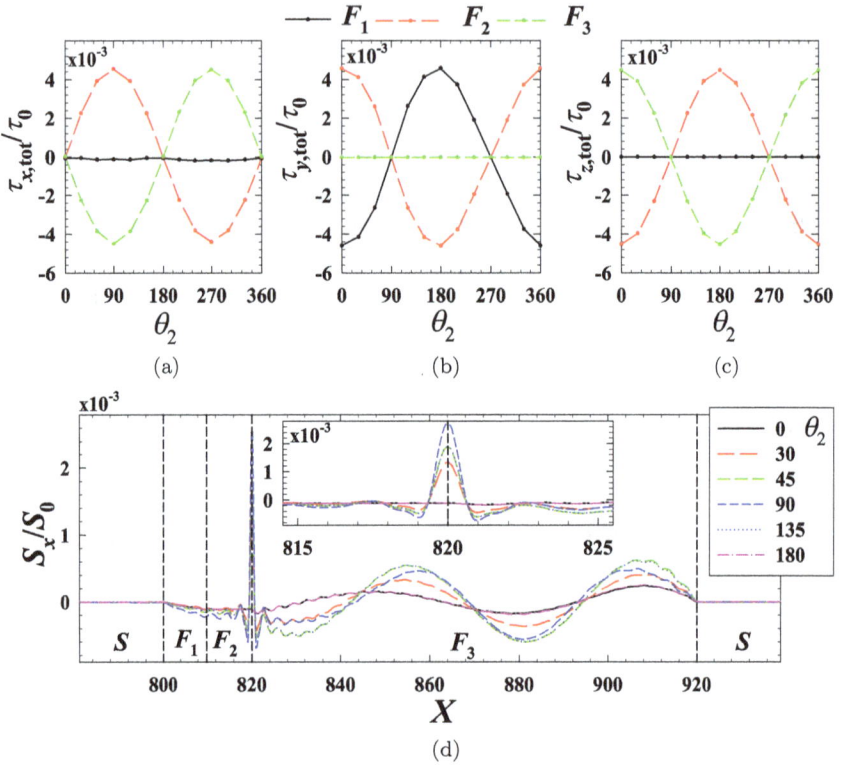

Fig. 11.11 Top row, panels (a)–(c): the three components of the integrated torque within each of the ferromagnet regions (overhead legend) of an $SF_1F_2F_3S$ junction as a function of the angle θ_2, see Fig. 11.1. The system parameters are those used in Fig. 11.6(a). The internal fields in F_1 and F_3 are along the z and y axes respectively. The sum of $\tau_{i,\text{tot}}$ ($i = x, y, x$) over all three ferromagnetic regions vanishes for each component. In the bottom panel, the spatial behavior of the normalized x-component of the spin current is shown throughout the system for a few select magnetization orientations θ_2 (see legend). The inset is a magnification of the region centered around the F_2/F_3 interface located at $X = 820$. Vertical dashed lines in the main plot mark interface locations.

we also observe that the total torque in either F_1 or F_3 nearly vanishes over the whole angular range of θ_2. This follows from the expression for the local torque, Eq. (9.6), which implies that $\boldsymbol{\tau}$ is orthogonal to the exchange field vector $\mathbf{h}$, and the magnetization $\mathbf{m}$. For example, considering the leftmost panel the only component of the exchange field in F_1 is along z, and since $\tau_x \sim m_y h_z$, a y component of the magnetization is needed in F_1 to generate a torque along x. However $\mathbf{h}$ in the adjacent F_2 rotates solely in the

$x-z$ plane, and thus m_y vanishes in F_1 except in a narrow region near the interface (see Chap. 9) resulting in a very small value for the averaged $\tau_{x,tot}$.

The local spin current density $\mathbf{S}$ is a local quantity, and when it is spatially nonuniform the resulting torque influences the magnetization configurations and the flow of spin-polarized currents. It is therefore insightful to examine also in this case, as we did in Fig. 11.10, and earlier, the spatial behavior of the spin current. The results are displayed in the bottom panel of Fig. 11.11. For clarity, only the x-component, S_x, is plotted since the other spin components behave similarly. A few representative angle orientations, θ_2, are included in the plot (see legend). We verify that S_x vanishes in the outer S electrodes, as it should. Within the larger F_3 region, the spin current undergoes regular oscillations which are much harder to distinguish in the narrow F_2 and F_1 regions. To compare with the previous results, we note from this panel that the change in the x-component of the spin current, ΔS_x, across the F_1 boundaries is negligible, in agreement with the results in the left top row panel. In F_2, this component of the current is very small near the left interface, but it increases near the right edge, so that ΔS_x agrees well with the $\tau_{x,\text{tot}}$ variations in F_2 observed in the left top panel. Also in agreement is the enhancement of the spin current S_x that is narrowly peaked at the F_2/F_3 interface for θ_2 values corresponding to near normal to the plane ($\theta_2 \approx 90°$). Finally, since the interface between F_3 and the right S electrode has $S_x = 0$, the change ΔS_x in the F_3 region arises entirely from the value of the spin current at the F_2/F_3 interface, thus resulting in ΔS_x that is exactly opposite to that in F_2.

11.4 Summary

In equilibrium, singlet Cooper pairs carry no net spin, hence in SF structures any spin current either flows only within the ferromagnets due to STT's generated by their exchange interaction, or it flows by means of induced equal-spin triplet correlations, where the Cooper pairs have a net spin component of $m = \pm 1$ on the spin quantization axis and they can reside in both S and F regions. We have extensively discussed in Chaps. 3 and 5 the generation of long-range triplet proximity effects in superconducting heterostructures with magnetic inhomogeneity: by introducing magnetic inhomogeneity, e.g., inclusion of an additional magnet with misaligned exchange field, the Hamiltonian no longer commutes with the total spin operator and equal-spin triplet correlations can then be induced. Equal-spin triplet correlations thus play a fundamental role in spin valve structures, as we saw in Chap. 5 and, as we have seen in this chapter and will see in the next one, in the physics of ferromagnetic Josephson junctions with inhomogeneity in the ferromagnet structure. As in spin valves, one of the simplest and cleanest ways to introduce controllable magnetic inhomogeneity in a Josephson junction, is through the insertion of bilayer or trilayer of uniformly magnetized ferromagnets. Experimentally, this has the advantages of reproducibility and easy manipulation of the relative exchange field orientation. These structures also provide direct evidence of triplet correlations as we saw in Chap. 6. The behavior of the triplet amplitude is anticorrelated (see Fig. 6.4 for a vivid example) to that of the critical temperature and associated singlet correlations, evidencing singlet to triplet conversion.

The oscillatory and long-range pair correlations lead, as we have seen in Sec. 11.3.2, to new behaviors in the current-phase relations (CPRs) in SFS junctions with a nontrivial magnetic structure. The CPR can in general contain not only the first harmonic but also higher order harmonics, i.e., $I(\Delta\varphi) \approx I_1 \sin(\Delta\varphi) + I_2 \sin(2\Delta\varphi) + \cdots$. In conventional Josephson junctions without any magnetic interactions, the magnitude of I_2 is much smaller than that of I_1. However, in SFS Josephson junctions, such as those we have studied in this chapter, the roles of the first harmonic and the second harmonic can be reversed, and the CPR can be largely dominated by the second harmonic. In this regime, both $\Delta\varphi = 0$ and $\Delta\varphi = \pi$ states can be stable or metastable, and they, as seen in Chap. 3, can coexist.

The effect of the interaction of the spin current with the magnetization in layered ferromagnetic junctions with multiple ferromagnets has important consequences for memory technologies. Indeed, storage of information bits depends on the precise relative orientation of the magnetizations in two F layers, where nonconserved spin currents reflect the mutual torque acting on the magnetic moments. The corresponding spin transfer torque (STT) can also switch magnetizations when a spin-polarized electrical current flows perpendicular to the layers. Spin-transfer torque is known to occur in a very broad variety of materials, making it an attractive switching mechanism. For equilibrium spin currents, tuning the supercurrent (via $\Delta\varphi$) directly influences the STT when varying the relative in-plane magnetization angle [35]. The direction of supercurrent flow however is not simply related to the direction of the induced torque that tends to align the magnetic moments. The triplet correlations generated in these Josephson junctions (with noncollinear relative magnetizations), can also induce spatial variations in the spin currents responsible for the mutual torques acting on the ferromagnets.

We have considered in this chapter the DC Josephson effect in structures of the $S(F)S$ type where the middle (F) symbol stands for two or three ferromagnetic layers, or two F layers separated by a normal spacer. In these ferromagnetic junctions, a Josephson current will flow in the absence of an applied voltage, if a phase difference between the two S layers exists. This is the case we have looked at, leaving some aspects of the voltage driven structures for Chap. 12. In these structures, the phase difference drives charge currents and spin currents as well, and we have seen how to study both.

For $SFFS$-type spin valves, we have studied how these charge and spin transport properties vary with in-plane relative magnetization angles. When a third F layer is present $(SFFFS)$, we have allowed, in this chapter, the central F region to have also an out-of-plane magnetic orientation component while those for the two outer F layers are still in-plane as usual. The self-consistent formalism ensures, as seen in Sec. 7.2, that the charge conservation law is satisfied and that the proper relations that balance the STTs and the gradients of the spin current, given in Eq. (9.5), hold. We have seen results for a wide range of values of the geometrical and orientation parameters, as well as reasonable interfacial scattering.

The examples given in this chapter were organized in several subsections. First, we have considered (Sec. 11.3.2) the current-phase relations (CPRs) as a function of geometrical parameters (layer thicknesses) at fixed relative angles between the in-plane magnetizations: parallel (P), antiparallel (AP), and normal (N). We find that in general, larger geometric asymmetry (the ratio of the thicknesses of the two outer F layers) leads to larger superharmonic (π periodic) behavior. This is particularly pronounced in the two-magnet case. We saw that the strength of interfacial scattering can affect the magnitude of the critical current. Next, the effects of magnetization misorientation on the CPR have been discussed. We have seen that these effects are profound. In particular, when sweeping the relative in-plane angle from P to AP at fixed phase differences $\Delta\varphi$ between two S electrodes, the supercurrent flow first vanishes at N configurations, and this is followed by reversal of its direction. This can be understood to a very large extent by noting that the generation of induced spin triplets (as discussed in Chap. 5) is correlated with magnetic inhomogeneity and hence it should be related, via this phenomenon, to the CPR relationships. Therefore one is led to discuss triplet generation, which is done in Sec. 11.3.4. There we also examined the local spatial behavior of both the

$m = 0$ and $m = \pm 1$ triplet correlations and carefully quantified each component as functions of magnetic orientations and $\Delta \varphi$. The results clearly demonstrate the existence and importance of the singlet-to-triplet conversion in these Josephson junctions. Finally, in Sec. 11.3.5 we have discussed spin transport and shown that, due to the interaction between the charge current and the magnetizations, both the spin current and the STT oscillate in the F regions. By varying $\Delta \varphi$ or the magnetic misalignment angles, the phase of their oscillations can change accordingly. In addition, we have seen that the methods used ensure that the spin current gradient equals the STT, as fundamental laws require.

One can reasonably hope that these results will guide the experimentalist in choosing optimal configurations for building devices such as low dissipation memory storage units, which are expected to rely on the behavior of the Josephson junctions studied here. We have seen that the flow direction of charge supercurrents can be controlled by varying the relative orientation angles in $SFFS$ and $SFFFS$ heterostructures. This property renders these systems prominent candidates in making superconducting spin valves. Adjusting the layer thicknesses can lead to a modified behavior of the CPR. In particular, experimentalists should consider ferromagnetic layers with asymmetric widths while fabricating such spin valves in order to maximize the switching effects.

In the next chapter, we will study some aspects of transport in voltage driven magnetic Josephson junctions.

Chapter 12

Ferromagnetic Josephson Junction Valve Structures: Quasiparticle Conductance

12.1 Basic Ideas

After having studied in Chap. 11 the static (zero applied voltage) properties of ferromagnetic Josephson junctions, we will now consider in this chapter aspects of bias driven transport for Josephson junctions made by adding an additional S layer to the outside of the spin valve structure whose thermodynamic and transport properties we have already extensively studied in Chap. 6 through Chap. 10. An example of such a structure is shown in Fig. 2.2. What results from adding this S layer is an example, and a rather intricate one, of a magnetic Josephson junction $S_1 F_1 N F_2 S_2$, some of whose properties we have just considered in Chap. 11, see for example Figs. 11.3 or 11.7. The focus will be, as in Chaps. 8 and 9, on the structure of the quasiparticle conductance G as a function of bias V, emphasizing the subgap region. In the numerical results shown in this chapter, taken from Ref. [37], material parameters appropriate to experimentally realized structures made using Nb, Co, and Cu will be emphasized.

The most prominent phenomena seen in Josephson structures are of course related to the Josephson current: the celebrated Josephson effects as described in elementary textbooks [3, 26]. In the DC effect, a phase-driven DC current runs through the Josephson junction at zero applied bias, up to a critical value, via the tunneling of the Cooper pairs. This is the effect we have examined in Chap. 11. In the AC effect an AC current driven by an applied bias occurs. The frequencies are in the GHz range for an applied bias of order 10 μeV [26]. But, in general, the Josephson current is not the only current that runs through a Josephson junction, as there is also the contribution from normal electron transport. Within a two-fluid model, one can express the net current in a Josephson structure [36] using the Resistively and Capacitively Shunted Josephson (RCSJ) model as:

$$I = I_c \sin\left(\theta\right) + I_{qp} + C\frac{dV}{dt}. \tag{12.1}$$

This equation describes a resistive and capacitive circuit element running in parallel with a pure Josephson junction of tunneling Cooper pairs. This is a non-linear equation and will result in hysteresis [36] in the current vs. voltage ($I - V$) curves if the time scale of the RC element, $\tau_{RC} = RC$, is greater than that of the Josephson junction $\tau_J = \Phi_o/2\pi I_c R$ (where Φ_0 is the superconducting flux quantum, R is the normal resistance of the junction and I_c the critical current). The I_{qp} term is called the quasiparticle current because that is what it is: it represents the contribution due to normal electron transport. Just as introduced in Chap. 7 it can be characterized via the bias dependence conductance,

$$G(V) \equiv dI_{qp}/dV. \tag{12.2}$$

One can measure the quasiparticle current by shunting the junction. This produces an $I - V$ characteristic at sufficiently small biases where the "capture" current is small (and the minimum nonzero voltage is also small). In the case of a non-tunnel junction, such as a clean or weak-link junction, there may exist unique subgap conductance features. A metallic weak-link is an $S(N)S$ structure in which the two S sides in the Josephson junction are separated by a thin metal, sometimes of the same material as the superconductor. For example, a point contact may be formed with one superconductor in contact with a superconducting substrate. Another example is the microbridge, where a thin bridge is etched between two superconducting "banks" [36]. Although the parts are continuously connected, the intermediate region in each case is a normal metal constriction. This is because the constriction is smaller than the coherence length which destroys superconductivity within the region. These constrictions are therefore studied in the dirty limit [1]. In the clean limit, the transport properties are not affected by a constriction or by impurity scattering [36]. In the cases studied in this book, the two S ends are separated by a combination of F and N materials: we are dealing with ferromagnetic junctions.

12.2 Methods

Therefore in this chapter we will study the quasiparticle current for $S_1F_1NF_2S_2$ structures, as characterized by the quasiparticle conductance. The same self-consistent methods which we discussed in Chap. 11 will be used. The conductances will be calculated as in Chap. 8. We will particularly emphasize the subgap structure of the quasiparticle current, as described by the bias dependence of the conductance $G(V)$. In Chaps. 8 and 9 we have studied the quasiparticle charge and spin transport in superconducting spin valve structures (F_1NF_2S). In these structures, as in those considered in Chap. 11, the presence of more than one ferromagnet allows for the formation of induced same-spin triplet Cooper pair correlations, which are long ranged. Due to this, and to the oscillatory nature of the singlet pair, we saw that the subgap features of the system are highly dependent on the magnetic misalignment angle ϕ and the thickness of the F_2 layer. In Chap. 10, we saw when studying G at or below the critical bias (CB), that the subgap conductance features are spin-split between contributions from incoming spin-up and spin-down electrons where, in this subgap region, one spin band features a peak in conductance while the other spin-band has a minimum. This leads to a peak conductance that oscillates with the dimensionless bias E between zero and the critical bias. We saw also that these conductance features are highly dependent on the interfacial scattering. We will see in this chapter that these dependencies also apply to the $S_1F_1NF_2S_2$ system. As in previous chapters, by G we mean the forward conductance, as opposed to any angularly averaged $\langle G \rangle$ seen in some examples such as in Fig. 8.6.

The methods needed for the self-consistent calculation of the pair potential are very close to those used in Sec. 11.2 of Chap. 11, which in turn are elaborations of those introduced in Chap. 2 and used also in other previous chapters. For the conductance calculations the procedures described in Chap. 8, recall in particular Sec. 8.3, will be used, with some needed wrinkles discussed below. The primary differences are due to the inclusion of the second superconducting layer. As discussed in Sec. 11.2 the form of the Hamiltonian is the same as was given in previous chapters. As in Chap. 11, because of the presence of two superconductors the pair potential is not necessarily real: there may exist a phase difference between the two S layers. With a single superconductor, there is only one phase associated with the s-wave symmetry and thus the pair potential can be taken to be real. Using the self-consistent method, as described in Secs. 2.5 and 11.2, one initializes the pair potential within each layer to a selected starting phase difference, $\Delta_1(y) = \Delta_0$ and $\Delta_2(y) = \Delta_0 e^{i\varphi}$ where Δ_1 and Δ_2 are the values of $\Delta(y)$ for y values within the S_1 and S_2 sides, respectively. In the present chapter the difficulties associated with maintaining a fixed phase difference between the two S layers that we have encountered in Sec. 11.2 obviously do not arise: the phase is allowed to iterate to its self-consistent value throughout the sample.

In the examples considered in this chapter it will be assumed that the two superconducting layers are made of the same superconducting material, and similarly that the two ferromagnetic layers are made of the same ferromagnetic material. This is realistic. As the phase of the complex pair potential is iterated when one repeatedly diagonalizes to self-consistency one finds that in equilibrium there are always two local stabilities in the phase: 0 and π. For an initial guess where the phase difference φ is not equal to 0 or π, the final self-consistent phase will always converge to the value which minimizes the free energy. This value may depend on the thickness (see Sec. 4.3) and on the relative magnetization angle of the ferromagnets. In non-self-consistent calculations (such as those done using the approximate analytic methods below), the 0 and π phases are degenerate, and need not be specified. In the numerical results in this chapter, taken from Ref. [37], the overall phase corresponding to the plot displayed is that which minimizes the free energy, i.e., the equilibrium phase.

As mentioned above, for the calculation of the conductance the transfer matrix method discussed in Sec. 8.3.2 to extract the quasiparticle conductance via Eq. (8.21) can be used, with some minor modifications. Specifically, one needs to properly allow for the fact that quasiparticles are driven into the system, so that incoming boundary conditions containing the proper reflection and transmission amplitudes, such as those in Eqs. 8.1 or 8.2, can be imposed. To do so, one has to take into account that the junction in Fig. 2.2 must be connected to some electrodes. Minimally, to perform the calculations, one needs to introduce a thin normal metal contact, which we will denote here by 'X', located to the left of the S_1 layer (see Fig. 2.2). This makes the structure studied effectively $XS_1F_1NF_2S_2$. This contact layer can be assumed to be thin enough to not affect the calculation of the pair potential through the proximity effect. For this reason, it is not necessary to include it in the equilibrium calculations. This contact layer, however, may possibly make an impact on the quasi-particle conductance. This impact may not be negligible, particularly as it introduces another interfacial barrier. Including this contact makes the calculation more realistic in terms of describing experimental setups, which of course must have contacts. It also makes some technical aspects of the calculation more tractable, as we shall see below. The contact region at the S edge is treated here completely self-consistently, and should not be confused with the source regions at the edge of the samples studied in Chap. 11. Here the current source is simply the applied voltage.

As in Sec. 8.2, one can then determine the reflection and transmission amplitudes by properly writing the incoming wavefunctions in terms of these coefficients. If a spin-up incoming electron in the leftmost layer is traveling in the normal metal contact X, the incoming wavefunction is:

$$\Psi_{X,\uparrow} \equiv \begin{pmatrix} e^{ik_N^+ y} + b_{\uparrow,\uparrow} e^{-ik_N^+ y} \\ b_{\downarrow,\uparrow} e^{-ik_N^+ y} \\ a_{\uparrow,\uparrow} e^{ik_N^- y} \\ a_{\downarrow,\uparrow} e^{ik_N^- y} \end{pmatrix} \tag{12.3}$$

and for a spin-down incoming electron:

$$\Psi_{X,\downarrow} \equiv \begin{pmatrix} b_{\uparrow,\downarrow} e^{-ik_N^+ y} \\ e^{ik_N^+ y} + b_{\downarrow,\downarrow} e^{-ik_N^+ y} \\ a_{\uparrow,\downarrow} e^{ik_N^- y} \\ a_{\downarrow,\downarrow} e^{ik_N^- y} \end{pmatrix} \tag{12.4}$$

where the second spin index of the reflection amplitudes denotes the spin of the incoming particle and $k_N^\pm = \left[1 \pm \epsilon - k_\perp^2\right]^{1/2}$ are the dimensionless normal metal wavenumbers. One can see the extreme similarity between Eqs. 12.3 (or (12.4)) and the corresponding Eqs. 8.1 (or (8.2)) although now the wavevectors are spin independent, rather than given by the expressions in Eq. (8.3).

One then proceeds as in Sec. 8.3.2 the only difference being that, since there are two S layers, the procedure associated with Eq. (8.13) and the self-consistent pair potential solution must be performed twice. But this is straightforward. Taking this into account, one can go on with the transfer iteration procedure which is described in Chap. 8, specifically in the paragraph that concludes with Eq. (8.20). The inclusion of the normal metal contact X means that the left-most layer can be described using incoming electrons and holes, as opposed to the electron-like and hole-like quasiparticles of a superconductor.

12.3 Resonance Phenomena

Before proceeding with the discussion of any numerical results, it is again pedagogically very appropriate to present, as we did in Sec. 10.2.1, non rigorous but analytic results which may lead to a better understanding of the underlying physics than that which is afforded by numerical results coming seemingly out of a black box. As in the previous case, these results exhibit resonance phenomena that help in the understanding of the subsequently presented numerical plots. These phenomena are much more easily understood from the analytic approximation.

To make an analytic calculation possible, one needs to make many approximations, very similar to those made in Sec. 10.2.1. A one-dimensional system and a pair potential constant in the S layers are assumed. Also, the thicknesses at the left and right ends will be taken to be infinite. Therefore, the only thickness dependencies come from the intermediate layers. The analytic calculation of the conductance is then a much simpler version of the numerical counterpart, which was just described above. The analytic solution leads to equations of the same form as Eq. (10.4). It does involve inverting multiple 8×8 matrices, which could in principle be done by hand but can be much more readily done using *Mathematica* or a similar symbolic package. However, the resulting expression is lengthy: the analytic form of the conductance is still complicated due to the sheer number of plane wave combinations of the coefficients that are present in each reflection amplitude. Therefore, it is better to perform an analysis in the same spirit as that in Sec. 10.2.1 by considering some of the possible relevant plane wave combinations, to extract periodic behavior. In Sec. 10.2.1 we found that the reflection amplitudes have a basic periodicity of $2\pi/h$ on the thickness of the F layer. This led to a periodicity of the conductance peak position of π/h, with the subgap peak conductance oscillating between the zero bias and critical bias for increasing thickness of the F layer. However, there is another plane wave combination to be considered here, which describes the sub-gap structure for $h = 0$. This we discuss next.

Consider an $N'NS$ system (a normal metal contact N' coupled to an N/S bilayer). If there is some interfacial scattering barrier at the N'/N interface, Andreev reflected holes from the N/S interface may interfere with the reflections at the N'/N boundary. We may look for resonance effects in the $N'NS$ system by examining the plane wave combination $e^{ik_N^+ D_N} e^{-ik_N^- D_N}$ at the critical bias $\epsilon = \Delta_0$. The wavenumber in the normal metal is then

$$k_N^{\pm} = [1 \pm \Delta_0]^{1/2} \approx 1 \pm \Delta_0/2 \tag{12.5}$$

The combination is in resonance when $e^{i\Delta_0 D_N} = e^{2\pi i n}$, where n is the integer corresponding to the nth resonant harmonic. Thus, the resonance in the amplitudes is expected to occur for

$$D_N = \frac{2\pi}{\Delta_0}n = \pi^2 n\Xi_0, \tag{12.6}$$

where the normalized pair potential Δ_0 is related to [3] the (dimensionless) coherence length Ξ_0 by $\Delta_0 = 2/(\pi\Xi_0)$. The conductance would be proportional to the absolute square of the amplitudes, thus the periodicity in the conductance peak resonance occurring at the critical bias should be

$$\Lambda_n = \frac{\pi^2}{2}n\Xi_0, \tag{12.7}$$

where Λ_n is dimensionless, in the same units as all capitalized lengths. It should not be confused with the mismatch parameter Λ defined in Table 2.1 which is unity in this chapter nor with the Ho structure wavelength of Chap. 5. The calculated conductance for varying thicknesses $D_N = \Lambda_n$ is easily found [37] to exhibit peaks with this Λ_n periodicity, which describes the formation of *new* peaks at the critical bias, shifting the previous n numbered peak into the subgap. This resonance is the result of multiple Andreev reflections, where an electron or hole is Andreev reflected off the S layer and is again reflected at the N'/N interface. For $h \neq 0$, which is the case of interest here, there is an additional oscillatory behavior due to the spin-split effect described in Chap. 10.

In a $S_1 N S_2$ Josephson structure, the additional S layer leads to additional dependencies of the conductance on the geometry. Electrons and holes Andreev reflected from the N/S_2 interface may be Andreev reflected again at the S_1/N interface. Including again a normal metal contact X with possible interfacial scattering at the X/S_1 interface, the quasiparticles which transmit through the S_1 layer may also reflect at the X/S_1 interface. The net result is a complex resonance effect that can be divided into two parts: resonance from reflections at the X/S_1 interface, and from reflections at the S_1/N boundary. I know of no simple explanation for the actual resonance behavior found in the analytic calculation and present here a purely phenomenological description. A resonance effect such as that described in Eq. (12.7), related to the thickness D_N, still occurs. There are then two harmonic resonance effects, labeled below as the even and odd harmonics. These depend on D_N. But in the odd harmonics there is also a dependence on D_{S1}. To describe it I introduce a term Q to take into account the actual analytically calculated (as explained) dependence of the resonance on D_{S1}. Thus:

$$\frac{\Lambda_{n,even}}{\Xi_0} = \frac{\pi^2}{2}n, \qquad\qquad n = 0, 2, 4, \ldots$$

$$\frac{\Lambda_{n,odd}}{\Xi_0} = \frac{\pi^2}{2}n - Q\left(\frac{D_{S_1}}{\Xi_0}\right), \quad n = 1, 3, 5, \ldots$$

(12.8)

where phenomenologically one finds that $Q\left(D_{S_1}/\Xi_0\right) \approx 1.2\ln\left(D_{S_1}/\Xi_0\right) + 1.94$ approximates the actual calculated resonance values. The even terms are due to reflections at the S_1/N interface and have the same form as Eq. (12.7) while the odd terms are due to reflections at the X/S_1 contact interface. The odd resonance values are reduced by the extra term Q which depends only on the ratio D_{S_1}/Ξ_0. However the limit where D_{S_1}/Ξ_0 is very large is obviously excluded, since then the phenomenological formula Eq. (12.8) makes no sense whatsoever in that limit, leading to negative resonance periodicities. The actual calculation in that case can still be performed analytically and does not lead to a peak in the subgap conductance.

Physically, these resonances are similar in nature to the spin-split oscillations found in the superconductor spin valves systems studied in Chap. 10. The wavefunctions of the electrons and holes reflecting off the F/S interface are combinations of plane waves with periodicities dependent on their spin-up or spin-down alignment. Andreev reflection is maximal when the wavefunctions of the opposite spin electron/hole pairs are in phase at the interface and minimal when they are out of phase. The peak resonances are also related to this phenomenon but in a simpler way: they arise from the multiple interference of the repeated reflections, which is why they occur when there are two scattering interfaces (one being with a superconductor). In an SNS system, the plane wave electrons/holes undergo multiple Andreev reflection (MAR, see Table 8.1 for acronyms), but are still oscillatory with thickness due to the what we may call self interference of the waves.

In the numerical results given in Sec. 12.4.2 the higher harmonics ($n \geq 1$) will be deemphasized. This is for two reasons: first, the peak positions are less well defined, as the pair potential is spatially dependent and its saturated value varies with layer thicknesses due to the proximity effect. Second, the $n = 1$ harmonic occurs for very large intermediate thicknesses: about five times the coherence length of the superconductor. The clean ballistic nanostructures built by experimentalists typically have a total intermediate thickness on the same order as the coherence length, often even less.

12.4 Quasiparticle Conductance Results

It is useful now to see the results one gets for G using the above procedures. First, I will show some approximate analytic results and then, in the next subsection, exact numerical ones, all taken from Ref. [37] which is the source of all the figures. The normalized ferromagnetic exchange field is always $h = 0.145$ for both ferromagnetic layers, and the dimensionless coherence length $\Xi_0 = 115$ for both superconducting layers. These are the values we have seen to be suitable to describe experimental samples using cobalt and niobium [21] in Chap. 6. Normalizations for G and E, as well as for all other quantities, are of course as in previous chapters: see Table 2.1 and the first paragraph in Sec. 7.1.2.

12.4.1 *Approximate analytic results*

Let us look first at examples of these approximate but analytic results for very simple $S_1 N S_2$ and $S_1 F S_2$ Josephson structures. In Fig. 12.1, we can

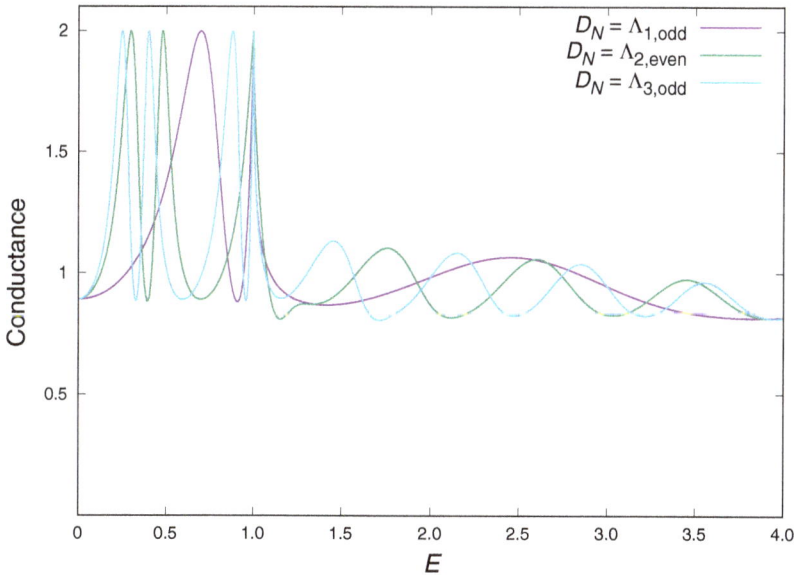

Fig. 12.1 Approximate analytic results for the quasiparticle conductance (G) vs. Bias (E) in an $S_1 N S_2$ structure with $D_{S1} = 180$ and several D_N thicknesses corresponding to resonant Λ_n values as given in Eq. (12.8). See text for discussion of the shifts in the "odd" peaks. The wavelength of the oscillations with E is seen to increase with Λ_n (i.e., with D_N). A single barrier at the X/S_1 contact with $H_B = 0.5$ is assumed.

see the conductance $G(E)$ for the S_1NS_2 structure with $D_{S1} = 180$ and resonance values of the N layer thickness as given by Eq. 12.8. The interfacial scattering at the X/S_1 contact is $H_B = 0.5$, a reasonable intermediate value. As discussed above, this Josephson structure has two sets of resonance values on D_N: the "even" and "odd" resonances. The odd resonances have an extra term $Q(D_{S1}/\Xi_0)$ (see Eq. 12.8) that decreases the resonance thickness value for the odd n harmonics. For the values included in this figure the decrease is substantial: from $\Lambda_1 \approx 568$ to 282. This split in resonances is due to the difference in the reflections at the S_1/N and the X/S_1 interfaces. The exact form of the additional Q term cannot be written in any compact closed form. It is phenomenologically obtained from analyzing results such as those shown in Fig. 12.1 as explained in the line below Eq. (12.8). In Fig. 12.1, a range of N layer thicknesses which encompass the first two odd resonances ($n = 1, 3$) as well as the $n = 2$ even resonance is included. As D_N increases, we see the shift of the critical bias peaks in $G(E)$ within the subgap region. However, due to the dual resonance structures, these peaks are not quite evenly spaced. Furthermore, the oscillations above the gap are not in phase and of course the wavelength is not directly proportional to the harmonic n for the odd resonances.

Consider now an S_1FS_2 example. In Fig. 12.2, the quasiparticle conductance computed in the analytic approximation is shown for a specific S_1FS_2 structure again with $D_{S1} = 180$. We have previously seen in Chap. 10 that the location of the conductance peak in E in spin valve structures oscillates between low biases and the CB as one varies the D_F thickness. This periodicity extends to all resonance peaks. Now, for this Josephson structure, within the approximations made, we see in Fig. 12.2 an oscillatory behavior with a total periodicity which is $2\pi/h \approx 43$ (for the h value used). This is due to the spin-split conductance, as in Chap. 10. In more complex structures one expects this periodicity to interfere with other periodicities present, leading to more complicated behavior. The periodic behavior will not be the same in three dimensional structures, not even in the quasi one-dimensional cases numerically studied below. The minimum F thickness in the plot is chosen to be $D_F = 85$ which is less than the first $n = 1$ resonance value ($\Lambda_{1,odd} \approx 282$). This is for two illustrative reasons: First, this value is the minimum total thickness of the intermediate layers (those between S_1 and S_2) in the numerical calculations on the $S_1F_1NF_2S_2$ ferromagnetic Josephson structure shown in the next subsection, so that comparisons are possible. Second, one wishes to show how the oscillations of the resonance peaks can shift a higher order harmonic peak into the subgap region, allowing for multiple subgap peaks. Indeed, in Fig. 12.2 we see a

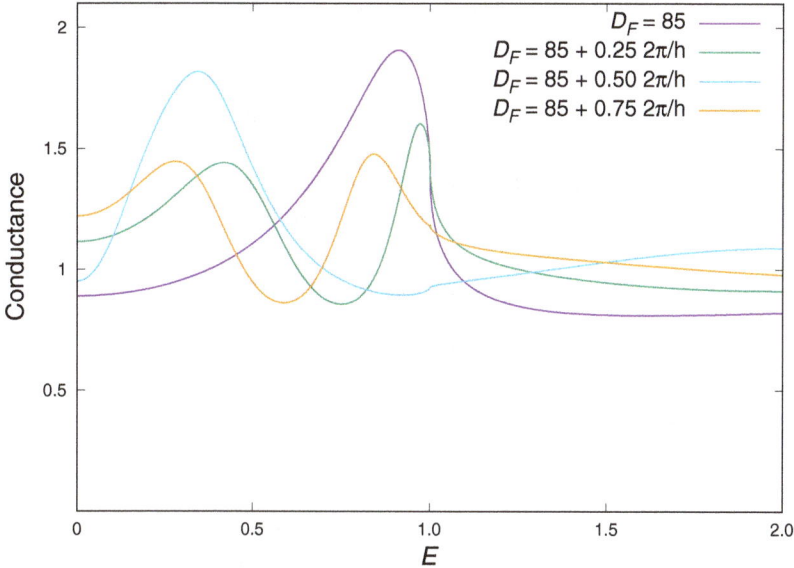

Fig. 12.2   Conductance G vs. bias in a $S_1 F S_2$ structure, in the analytic approximation, for four values of D_F (see legend). The location of the lowest E peak decreases with increasing D_F. G is plotted for one full oscillation of the thickness periodicity $2\pi/h \approx 43$. A single barrier at the X/S_1 contact, with $H_B = 0.5$ is included.

single conductance peak at $D_F = 85$. But as D_F increases, the peak splits into two subgap peaks (with one being very near the CB). Then, the two peaks reform at $D_F = 85 + \pi/h$ (at the half period mark) into a single subgap peak. Increasing D_F further, this peak splits again into two subgap peaks with one being now at very low biases. The second superconductor is needed to observe this: it is not seen in NFS structures such as those in Sec. 10.2.1, where the main periodicity is halved.

Therefore, this approximate analytic study shows that there are two sources of resonance in SFS Josephson structures that give rise to conductance peaks in the subgap region: one related to the field, the other to geometrical scattering. Furthermore, these peaks are oscillatory with varying thickness of the F layer. This is as far as we can go with this approximate procedure and we now have to turn to the full numerical method.

12.4.2   *Numerical results*

Having gained these insights, we turn now to numerical self-consistent results for the ferromagnetic Josephson structure $S_1 F_1 N F_2 S_2$. Recall that a thin normal metal contact X is also included, as explained in Sec. 12.1, the structure being then $X/F_1NF_2S_2$. The above analytic results of Sec. 12.4.1 will help guide us in the analysis, while keeping in mind that there are important differences: among them is that the numerical results are for three dimensional systems with a more complicated structure. Even more important, in the numerical calculations the pair potential is calculated self-consistently so that within each superconductor it is a function of position. The pair amplitude within the whole multilayer has a nontrivial position dependence. Furthermore, in the numerical calculation all layers are realistically finite in width.

In the remaining figures in this Chapter, all layer thicknesses are constant except that of the F_2 layer. The other thicknesses are in all cases $D_{S1} = D_{S2} = 180$, $D_{F1} = 30$ and $D_N = 40$. The normal metal electrode thickness is set to $D_X = 5$. Again, the results presented are for $h = 0.145$ and $\Xi_0 = 115$, for the reasons, related to experiment, Chap. 6, as mentioned above. We will also study the dependence of the quasiparticle conductance $G(E)$ on the X/S_1, the F/N and the F/S interfacial scattering. Note again that the results are for the voltage driven quasiparticle current. They do not reflect the zero bias current driven by a phase difference: the ordinary dc Josephson effect. The quasiparticle conductance is measured as explained in Sec. 12.1 — see remarks below Eq. (12.2). Therefore, in interpreting the plots, it should be kept in mind that the ultra-low bias conductance may be inaccessible in experiment, even with a hysteresis current from a shunted Josephson circuit,

12.4.2.1 *Geometry dependence*

In the next two figures, Figs. 12.3 and 12.4, we examine the dependence of the quasiparticle conductance features on D_{F2}. The range of D_{F2} values considered includes, as we shall see, an experimentally relevant region where one complete cycle of oscillation in the results occurs. The analysis is therefore a little less extensive than that done in Chap. 10 but it will suffice. In these two figures, the internal fields of the ferromagnetic layers are assumed to be parallel ($\phi = 0$). Below, in Sec. 12.4.2.2, we will study also the dependence of G on the misalignment angle ϕ at fixed D_{F2}. The interfacial scattering strengths $H_{B,i}$ are indexed from the far left X/S_1 to the right F_2/S_2 starting from zero ($0 \leq i \leq 4$): thus the X/S_1 barrier strength is $H_{B,0}$, the F/N interfacial barrier strengths are $H_{B,2}$ and $H_{B,3}$, and the F/S barrier strengths are $H_{B,1}$ and $H_{B,4}$. This notation is similar

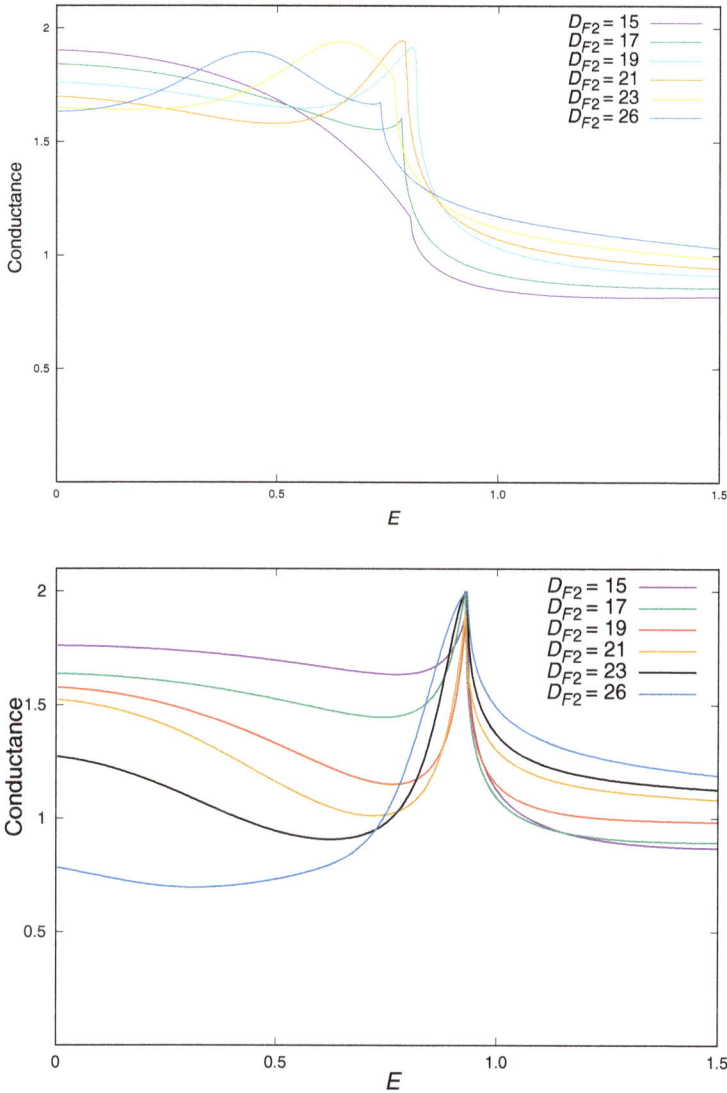

Fig. 12.3 Numerical results for the quasiparticle conductance G vs. bias E in a $S_1 F_1 N F_2 S_2$ structure for varying D_{F2} and $\phi = 0$. In the top panel, there is scattering at the F/N interfaces: $H_{B,0} = H_{B,1} = H_{B,4} = 0$, $H_{B,2} = H_{B,3} = 0.5$ (see text for notation). At the bottom panel, there are instead barriers at the F/S interfaces $H_{B,0} = H_{B,2} = H_{B,3} = 0$, $H_{B,1} = H_{B,4} = 0.5$. The values of D_{F2} are indicated in the legends. In both panels G at large E increases with D_{F2}.

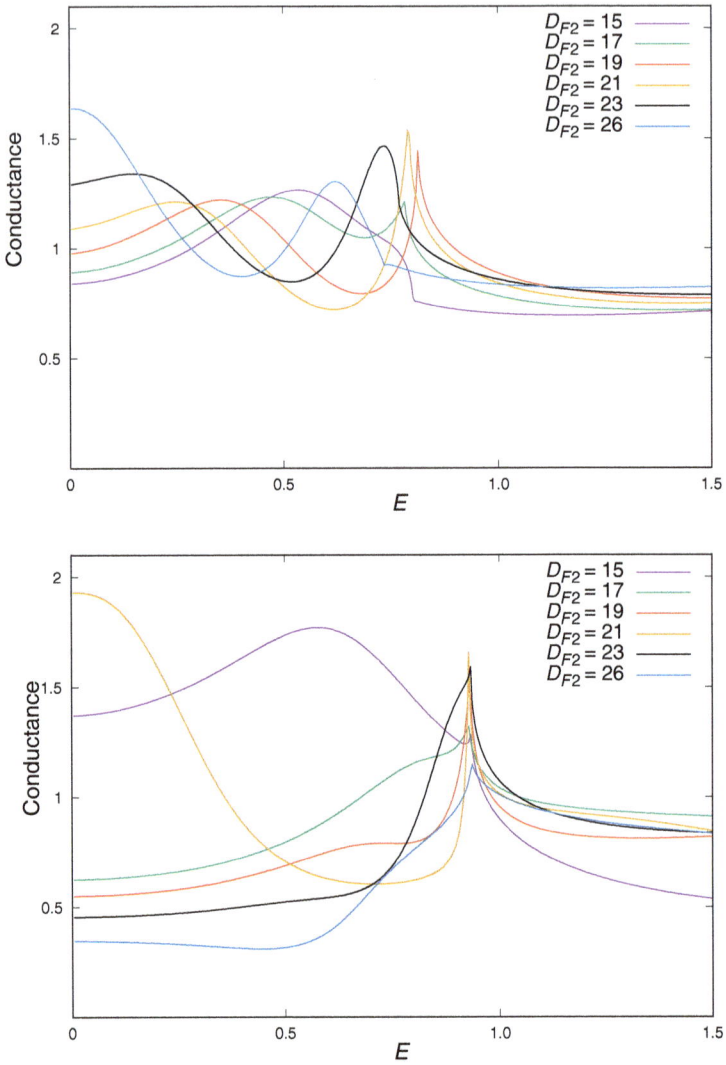

Fig. 12.4 As in Fig. 12.3, but now with $H_{B,0} = 0.5$. In the top panel there is also scattering at the F/N interfaces: $H_{B,1} = H_{B,4} = 0$, $H_{B,0} = H_{B,2} = H_{B,3} = 0.5$. In the bottom panel the scattering occurs at the F/S interfaces: $H_{B,2} = H_{B,3} = 0$, $H_{B,0} = H_{B,1} = H_{B,4} = 0.5$. In the top panel $G(E = 0)$ increases with D_{F2}. In the bottom one, $G(E \approx 0.6)$ decreases with the same variable.

to that used in Chap. 8 (see in particular Eq. (8.31)). The barrier strengths are either zero or 0.5. This intermediate value is a realistic one in experimental, Chap. 6 situations. The situation where all interfaces are perfect, i.e., $H_{B,i} = 0$, need not be considered, as it would be quite unrealistic experimentally. However, we will see that the overall interfacial effects are dominated by one set of material interfaces over the other. Therefore one can simplify matters by considering only cases where only the most relevant of the interfacial parameters are nonzero.

In Fig. 12.3, we see results for the bias dependence of the conductance. In the top panel, the results plotted are for F/N interfacial scattering, $H_{B,2} = H_{B,3} = 0.5$, while in the bottom one there is instead F/S interfacial scattering, $H_{B,1} = H_{B,4} = 0.5$. The X/S_1 interface is transparent ($H_{B,0} = 0$). Conversely, in Fig. 12.4 we can see results which include the effect of interfacial scattering at the X/S_1 contact interface, $H_{B,0} = 0.5$. There are again two plots: the top one includes F/N interfacial scattering, and the bottom one F/S interfacial scattering. These two figures combined illustrate also the influence of the X electrode, which should arise chiefly from the influence of the additional barrier. In both of these figures the forward conductance for $\phi = 0$ is plotted for several values of D_{F2} indicated in legend and caption, chosen to illustrate the oscillatory behavior with this variable. In Fig. 12.3, we can see in the subgap region a single peak structure: the peak of the conductance can be seen to shift from low bias to the critical bias over the range of D_{F2} shown. This single peak structure is similar to that studied in the figures shown in Sec. 10.2.2.1. There, the bias location of a subgap peak was shown to oscillate with the ferromagnetic layer thickness in F_1NF_2S spin valve structures which also had interfacial scattering at the F/N interfaces. The periodicity of the oscillations was shown to be π/h. When the peak is located between zero bias and the critical bias, there is a cusp feature in the critical bias conductance. In the bottom panel of Fig. 12.3, where the conductance is plotted for the same cases but for nonzero F/S scattering, we see a very different phenomenon: there is now a subgap minimum as opposed to a subgap peak. There is a marked peak at the critical bias. The subgap minimum is also dependent on D_{F2}: we see that the subgap minimum goes from being near the critical bias at $D_{F2} = 15$ to being near zero bias at $D_{F2} = 26$. This thickness dependence is different from that in the oscillatory peak structure in the top panel of Fig. 12.3; the periodicity of the minima is not π/h. Such minima, or dip structures have been discussed on and off in the literature of ordinary SNS Josephson junctions, usually described in terms of a subgap peak structure in the *resistance*, related to interfacial scattering at the S/N interfaces. This would translate to dips in the conductance.

In Fig. 12.4 the conductance plots are for the same situation as in Fig. 12.3 except that there is a nonzero barrier at the X/S_1 interface — $H_{B,0} = 0.5$. Besides comparing these plots with the numerical results of Fig. 12.3 as discussed above, it is perhaps more important that they should be compared with the analytic results presented above, such as those in Fig. 12.2. Specifically, we see that despite the stark differences that I have pointed out between the assumptions in the analytic and the numerical calculations, these numerical results are still in many ways qualitatively similar to the analytic ones. When D_{F2} changes in the range, chosen for the reasons discussed above, from $D_{F2} = 15$ to $D_{F2} = 26$, the behavior of G with bias in the subgap region changes, as seen in the top panel, from a single peak near the CB to a two-peak structure with one peak at low bias and one just below the critical bias. As D_{F2} increases from 15, we see the single peak shift into the subgap region until a second peak forms at the CB, at around $D_{F2} = 21$. The thickness difference between $D_{F2} = 15$ and $D_{F2} = 26$ is about $\pi/2h$, which is one quarter of the total main oscillatory pattern. Comparing this top panel with Fig. 12.2 where D_F varies, with X/S_1 interfacial scattering being present, we can see the same behavior going from a single peak to two distinct peaks as D_F is increased by $\pi/2h$. We can conclude that this means that the presence of electrode interfacial scattering doubles the periodicity in both cases, from $\pi/2h$ to π/h. This behavior is due to the reflections at the X/S_1 interface which form a second resonance effect.

Let us turn now to the analysis of the interfacial scattering effects. By comparing the top and bottom panels of Fig. 12.4, we see that a large change occurs when there are F/S scattering barriers. The subgap structure becomes much more complicated: there is no subgap peak, except at $D_{F2} = 15$ and $D_{F2} = 21$, but there is a noticeable inflection point within the subgap region at the other thicknesses. One can ascribe this to the combined effect of the usual peak structure and the dip, or minima structure seen in Fig. 12.3. The presence of the F/S barriers gives rise to dips, while the X/S_1 barrier provides a peak resonance in the MAR. These two effects do *not* share the same periodicity with the thickness: therefore interference phenomena complicate the overall effect, as discussed above. From this we can conclude that the F/S barriers have the most impact on the subgap structure. The electrode barrier is also important, however, as it leads to a resonance effect in the MAR which can lead to a second conductance peak or a complex inflection structure within the subgap, depending on what other barriers are in play.

12.4.2.2 *Superconducting valve effects*

Much of the interest in the spin-valve Josephson structure arises from its putative capability to store information in the relative orientation of the magnetization in the F layers. As we have repeatedly emphasized in this book, the angular dependence of the conductance constitutes a valve effect in the system. In the superconducting structure ($FNFS$) studied in Chap. 10, we saw (particularly in Sec. 10.2.2.1) that a large valve effect, which was actually a spin valve effect, exists in the subgap conductance for certain thicknesses of the F_2 layer. In this subsection we want now to discuss the determination of the angular dependence of G for the $S_1 F_1 N F_2 S_2$ structure and the viability of the valve effect found.

To do this, let us analyze G for two of the thicknesses plotted in Figs. 12.3, and 12.4, $D_{F2} = 15$ and $D_{F2} = 26$. The values of the other thicknesses are kept fixed to the same values as before. These two thickness values were chosen because they are separated by a value of $\pi/2h \approx 11$, one quarter of the basic periodicity with $H_{B,0} = 0.5$ and half the basic periodicity for $H_{B,0} = 0$. We wish then to show how to relate the dependence of G on the angle ϕ with the spatial dependence found in Figs. 12.3, and 12.4. For each of these two thicknesses, let us look at each of the two cases shown in both panels of these previous figures. Within each figure one can compare the effects that the other interfacial barriers have on the angular dependence of the conductance.

In Figs. 12.5 and 12.6 there are plots of the forward conductance for $D_{F2} = 15$, which display its dependence on the misalignment angle ϕ. The $\phi = 0$ results are the same as those shown for the same thickness in Figs. 12.3 and 12.4 respectively, since except for the ϕ variation, the situation is the same. Thus, in Fig. 12.5 the conductance is plotted for $H_{B,2} = H_{B,3} = 0.5$ in the top panel and for $H_{B,1} = H_{B,4} = 0.5$ in the bottom one, all other interfaces being transparent. We see then that in the top panel, the peak in the conductance at low bias for $\phi = 0$ transitions to a subgap peak at $\phi = 180°$. This is reminiscent of the thickness dependence, where the single low bias peak at $D_{F2} = 15$ transitions into a subgap peak at $D_{F2} = 26$, as seen in Fig. 12.3 (top panel). This analogy extends to the $H_{B,1} = H_{B,4} = 0.5$ case in the bottom panel of Fig. 12.5. At $\phi = 0$ there is only a small dip in the conductance near the critical bias while

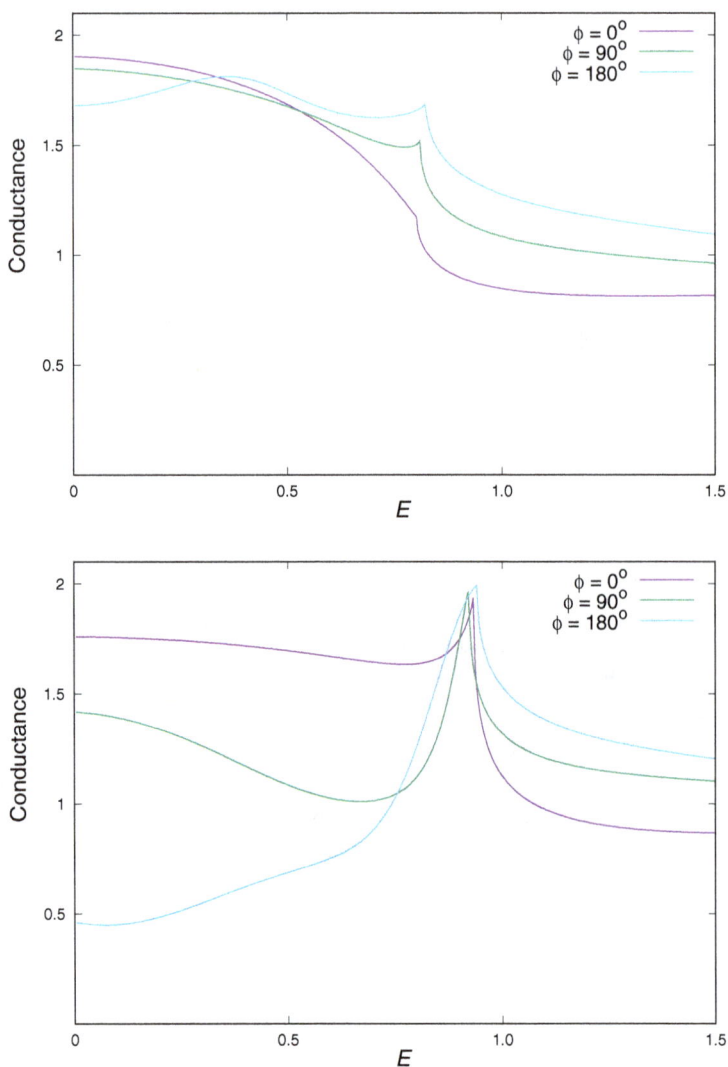

Fig. 12.5 Numerical results for the conductance G vs. bias E in a $S_1 F_1 N F_2 S_2$ structure with $D_{F2} = 15$, for three values of ϕ (see legend: G increases with ϕ at large E). In the top panel there is scattering at the F/N interfaces, $H_{B,0} = H_{B,1} = H_{B,4} = 0$, $H_{B,2} = H_{B,3} = 0.5$. In the bottom panel there is instead scattering at the F/S interfaces, $H_{B,0} = H_{B,2} = H_{B,3} = 0$, $H_{B,1} = H_{B,4} = 0.5$. The $\phi = 0$ curves are the same as the $D_{F2} = 15$ plots in Fig. 12.3.

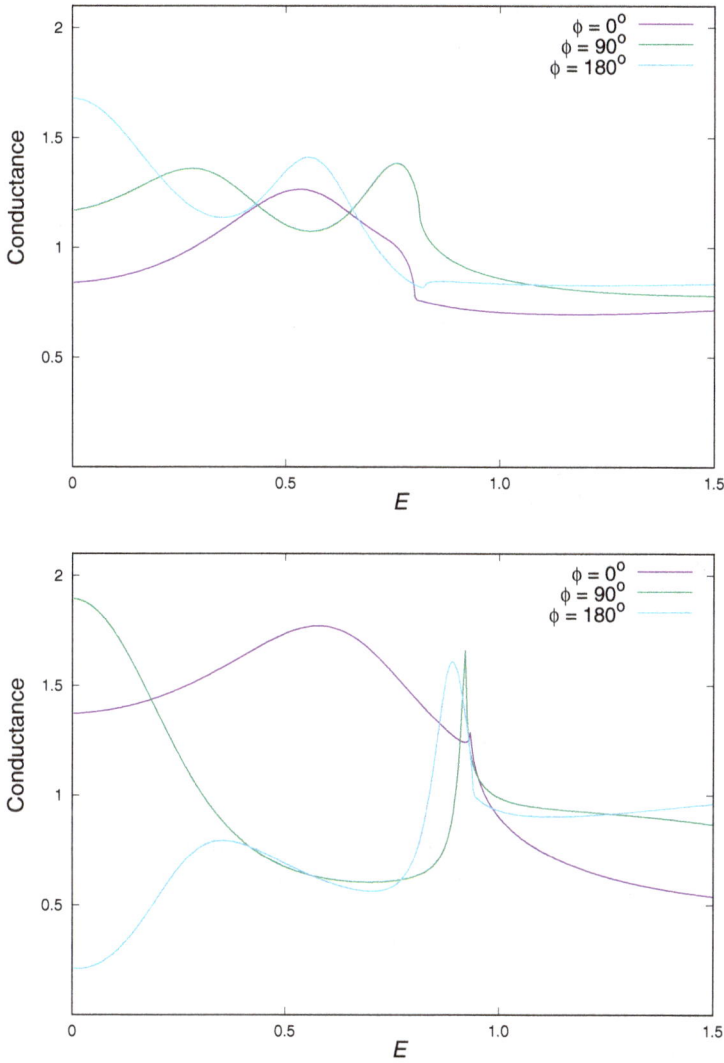

Fig. 12.6 Numerical results for G vs. E at three values of ϕ. Everything is as in Fig. 12.5 except the interfacial scattering. There is now X/S_1 scattering with $H_{B,0} = 0.5$ in both panels. In addition, in the top panel there is also interfacial scattering at the F/N interfaces: $H_{B,1} = H_{B,4} = 0$, $H_{B,2} = H_{B,3} = 0.5$, while in the bottom panel, there is scattering at the F/S interfaces: $H_{B,2} = H_{B,3} = 0$, $HH_{B,1} = H_{B,4} = 0.5$. The $\phi = 0$ curves are the same as the $D_{F2} = 15$ plots in Fig. 12.4.

at $\phi = 180°$ the low bias conductance drops to a minimum value. This is similar to the changes from $D_{F2} = 15$ to $D_{F2} = 26$ in the bottom plot of Fig. 12.3. However, in addition to the dip structure, there appears to be a small inflection near $E = 0.4$. This is not seen in the corresponding case of Fig. 12.3. This feature is more reminiscent of those found in the bottom panel of Fig. 12.4.

The organization of the plots in Fig. 12.6 is the same as in the previous figure, and the relation of this figure to Fig. 12.4 is analogous to that between Figs. 12.5 and 12.3. In the top panel of Fig. 12.6 there is, qualitatively, a single-peaked conductance at $\phi = 0$. At $\phi = 90°$ the single peak splits into a subgap peak and a CB peak. Then, at $\phi = 180°$, the conductance has two subgap peaks, one at low bias and one just below the critical bias. This angular dependence is also qualitatively the same as the thickness dependence going from $D_{F2} = 15$ to $D_{F2} = 26$, as in Fig. 12.4. However, unlike in the superconducting spin valve case discussed in Chap. 10, the introduction of additional barriers does not enhance the valve effect. The peaks decrease in value with increased F/N barrier. If we consider the situation when F/S barriers are present instead, as in the bottom panel of Fig. 12.6, we see a very different angular dependence from that of the other panel. The similarities with the D_{F2} dependence remain, in that there is a complex angular dependence with both a peak and dip structure. However, at $\phi = 180°$, the inflection point now leads to the formation of a small peak at a bias value substantially lower than the critical bias, while a minimum near zero bias remains.

In general, we can conclude that the angular dependence between $\phi = 0$ and $\phi = 180°$ at fixed D_{F2} is similar to the D_{F2} dependence as this latter quantity goes from $D_{F2} = 15$ to $D_{F2} = 26$ for the parallel configuration, $\phi = 0$. This is a striking result: in the superconducting spin valve we have studied in Chap. 10 the angular dependence consists of a uniformly increasing or decreasing conductance peak as one goes from the parallel to the antiparallel configuration, with the position of the peak being dependent on D_{F2} only. In the ferromagnetic Josephson structure, on the other hand, the angular dependence does not affect the height of the peaks, but it affects instead the position of the peaks just as with the D_{F2} dependence. This allows for an extremely large superconducting valve effect for almost *any* bias value, as can be seen for example in Fig. 12.6 (bottom). There we can see that there is a difference in conductance on the order unity (that is, one quantum of conductance per channel), between $0°$ and $90°$ at low biases, and between 0 and $180°$ closer to the CB. Just as with the thickness dependence, the F/S barriers have the greatest impact on the angular dependence of the system. Of course, in considering the similarity between the ϕ and the D_{F2} dependencies one should never forget that changing the former is much easier than changing the latter.

In Figs. 12.7 and 12.8 the same results are shown as in Figs. 12.5 and 12.6 respectively, and with the same parameter values, except that now $D_{F2} = 26$, the larger "end of cycle" value shown in Figs. 12.3 and 12.4. In the $H_{B,0} = 0$ case shown in Fig. 12.7, one sees in both panels that the angular dependence closely resembles that in the corresponding plots of Fig. 12.5 *but for supplementary angles*. That is, the $G(E)$ curve for $\phi = 0$ and $D_{F2} = 15$ is similar to that for $\phi = 180$ and $D_{F2} = 26$ and vice versa. This is clearly related to the π/h periodicity: the thickness difference between 15 and 26 represents half a period in the D_{F2} dependence as I have already mentioned. Between the parallel ($\phi = 0$) and antiparallel ($\phi = 180°$) configurations, the value of ϕ advances by π, half a period, and oscillations with ϕ are closely correlated to those with D_{F2} at fixed angle.

Similarly, Figs. 12.8 and 12.6 again differ only on the value of D_{F2} and their similarities and differences deserve notice. The similarities, in most cases, are analogous to those discussed in the previous paragraph, but there is an obvious difference in the bottom panels, where there is interfacial scattering at both of the S/F interfaces. In that case the behavior of G below the CB at the larger value of D_{F2} (Fig. 12.8) is more reminiscent of a tunneling behavior, as compared with the bottom panel of Fig. 12.6. The other cases are less affected qualitatively. It is perhaps true that given the persistent oscillatory behavior with geometry of all quantities involved, one should not put too much store on comparing just two thicknesses. Nevertheless, the relation between the ϕ and D_{F2} behaviors exists and should be kept in mind. One must also recall that at the value of h studied here, chosen for reasons related to experiment, is relatively small compared to unity: as a result the proximity length for the singlet pairs is comparable to the values of D_{F2} and $D_{F1} = 30$. At much larger values of these thicknesses the difference between the proximity lengths for singlet and odd triplet pairs would come into play even more strongly and perhaps produce a more marked difference between the behavior at $\phi = 0$ or $\phi = \pi$ and the other angles.

At larger bias values the difference between the $\phi = 0$ and $\phi = \pi$ behaviors seems at first somewhat counterintuitive, as it is the opposite of that seen in ordinary spin valves. This is due to the proximity effect of the superconductor which allows for the "spin-split conductance", discussed in Sec. 10.2.2.1, where the dominance of one spin-band conductance over

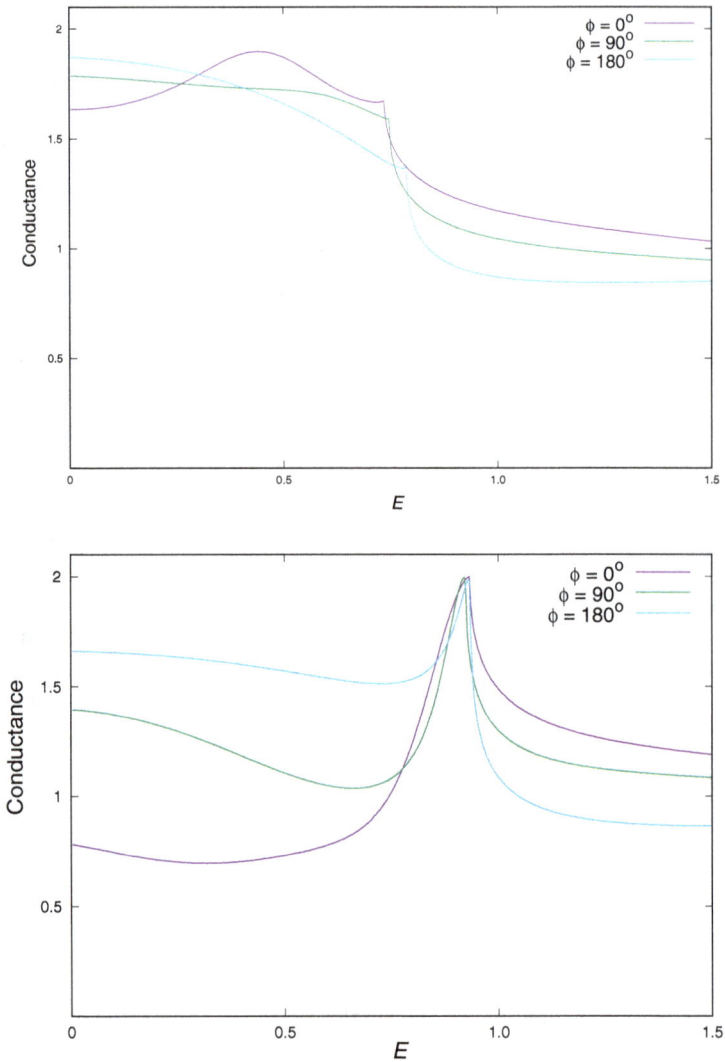

Fig. 12.7 G vs. E plots for an $S_1F_1NF_2S_2$ structure for three values of ϕ. All parameters and arrangements are as in Fig. 12.5 except that now $D_{F2} = 26$. Notice that now the values of G at large E decrease with ϕ, instead of increasing.

the other will periodically shift with the F layer thickness. Figures from Fig. 10.5 through Fig. 10.10 amply demonstrate this effect.

The patterns in the angular dependence are similar in all of the barrier cases shown. As an additional instance, in the conductance for $\phi = 0$ in

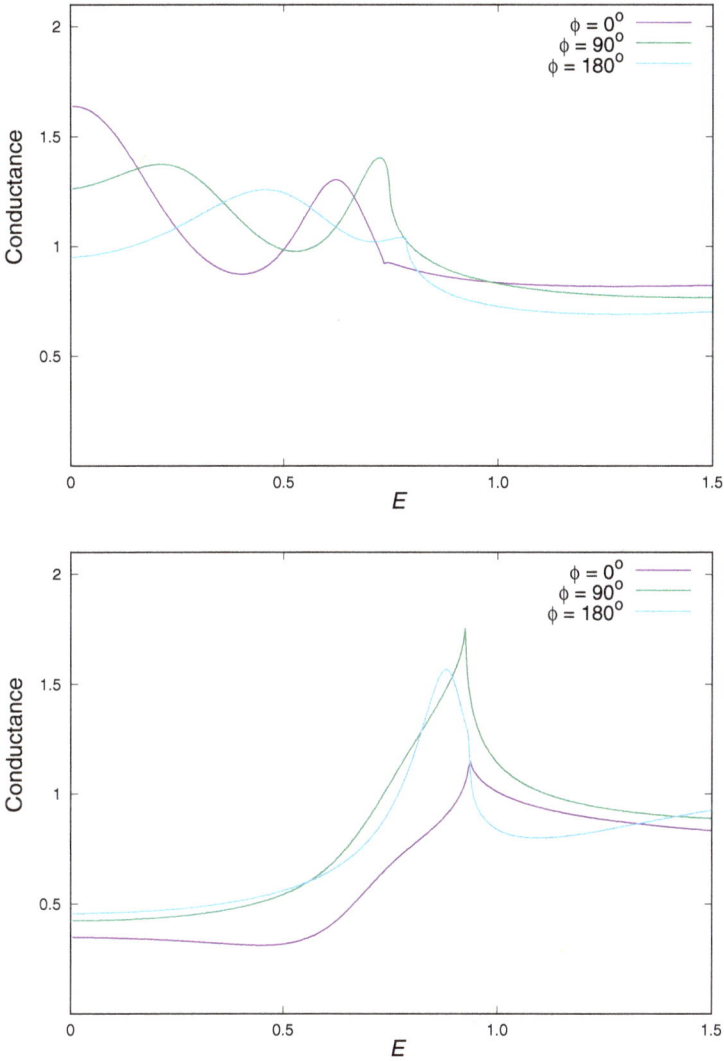

Fig. 12.8 G vs. E in an $S_1F_1NF_2S_2$ structure for three values of ϕ. All settings are as in Fig. 12.6 but for $D_{F2} = 26$ instead of $D_{F2} = 15$. In the top panel $G(E = 0)$ decreases with ϕ, in the bottom panel it increases slightly.

the top panel of Fig. 12.8 (which is the same as plotted in the same panel of Fig. 12.3 at $D_{F2} = 26$) one can see two peaks: one at low bias and one within the subgap region. For $\phi = 90°$ now the two peaks shift to the right, with the higher E peak moving into the critical bias region. Finally at $\phi = 180°$ these conductance peaks have merged into a single feature just below the critical bias. This is the same behavior as seen in Fig. 12.1, except that the periodicity in the thickness is $2\pi/h$. Indeed, the angle ϕ advances the phase of the overall oscillatory spin-split behavior by $\pi/2$ when going from a parallel to an antiparallel configuration. In the $FNFS$ case, only for certain ranges of thicknesses would the valve effect be noticeable (when the peak was in the middle of the subgap region). We see now that in the $SFNFS$ structure, the peaks change in position with ϕ. This means the valve effect is apparent for any thickness, as any minimum found at $\phi = 0$ will become a maximum when the magnetization is rotated by a certain angle ϕ. We can also note that external electrode scattering doubles the effective periodicity in the conductance subgap features, while the angular dependence between parallel and antiparallel advances the phase of the thickness by the equivalent π/h wavelength in both cases.

12.5 Summary

In the main part of this chapter we have considered the bias dependence of the quasiparticle conductance of $S_1F_1NF_2S_2$ ferromagnetic Josephson structures. These results were obtained via the numerical self-consistent approach which has been described and used throughout the book.

To better understand these numerical results, I have introduced, as in Chap. 10, an approximate but analytic approximation within which an understanding of resonance and other oscillatory phenomena can be sought. This was developed in Sec. 12.3. In this approximation, a one-dimensional multilayer with a constant pair potential Δ_0, in the S layers, was assumed. A normal metal contact (with possible interfacial scattering) was included. I first considered an $N'NS$ system. Then, for sufficiently large thicknesses D_N the conductance forms "resonance" peaks. These form, at equally spaced intervals Λ_n which I called the resonance thicknesses, with harmonic number n. These are given in Eq. (12.7). For higher ordered harmonics ($n \geq 1$) there exist multiple peaks which are also evenly spaced between the zero bias and critical bias conductance. These resonances are due to the interference of Andreev reflected particles at the several interfaces. At higher harmonics ($n \geq 1$) the conductance is oscillatory just above the critical bias and slowly decays at the same rate for all harmonics. The frequencies of these oscillations are approximately proportional to the harmonic number n.

For an S_1NS_2 case the same approximate method shows that there are two resonance behaviors: "even" and "odd" (see Eq. (12.8)). The even harmonic resonances are equally spaced, but the odd harmonics have an additional term that depends on the ratio of D_{S1}/Ξ_0. This term substantially reduces the resonant thickness of the odd conductance peaks. The oscillatory conductance above the gap is also shifted by this D_{S_1} thickness dependence in the odd harmonic thicknesses.

The analytic approximation was also applied to the calculation of $G(E)$ in SFS as well as SNS systems. Recall that in Chap. 10 we studied, both in a similar analytic approximation (Fig. 10.1) and numerically (e.g., Fig. 10.4), the spin-split conductance of NFS trilayers, and found that the bias location of the conductance peak oscillates, for varying thicknesses of the F layer over a wavelength of π/h. In the present chapter we saw similar effects (see Fig. 12.2) on the $n = 1$ harmonic for an $S_1 F S_2$ trilayer, where there are two conductance peaks, both oscillating in position together between the subgap region and the zero bias conductance for the low bias peak, and the critical bias and subgap region for the higher bias one. Between the two thickness values marking one period, each peak splits, resulting in multiple subgap peaks in the conductance. In this analysis of these Josephson structures, we saw that for even relatively small values of the F layer thickness (less than the coherence length of the superconductor), the conductance may display multiple subgap peaks. This is because spin-split oscillations can pull the higher order harmonic peaks into the subgap region since the first harmonic thickness ($n = 1$) is reduced by the presence of the S_1 layer.

Armed with this qualitative, approximate but physical understanding of the F layer thickness dependence of the SFS analytic calculation, we were then ready to consider the results for the fully self-consistent $S_1 F_1 N F_2 S_2$ ferromagnetic Josephson structure. In Sec. 12.4.2.1 the F_2 thickness dependence in the parallel configuration of the F layer magnetizations ($\phi = 0$) was emphasized (see Figs. 12.5 and 12.6). In these numerical results we saw the same qualitative periodic features of the subgap conductance as found analytically for the SFS system.

We concluded Sec. 12.4.2 with a subsection, Sec. 12.4.2.2 on the valve effect, via the the ϕ dependence of G in the ferromagnetic Josephson structure. We saw in Figs. 12.7 and 12.8 results for the conductance for multiple angles ϕ of the relative orientation of the ferromagnetic layer magnetizations for a $S_1 F_1 N F_2 S$ structure. The angular dependence is surprisingly similar to that on D_{F2}. By rotating ϕ between the parallel and antiparallel configuration, the phase of the spin-split conductance oscillations advances by $\pi/2$ in the $H_{B,0} \neq 0$ case and by π in the $H_{B,0} = 0$ case. In both cases, this is approximately equivalent to increasing the thickness by π/h. Also in both cases, this is in stark contrast to what happens in the F_1NF_2S structure, where (see Chap. 10) the angular dependence is found only in the subgap peak height and *not* in the position of the peaks within the subgap. This means that in ferromagnetic Josephson junctions there is a very large valve effect, on the order of the quantum of conductance per channel, which may prove useful in future spintronic devices.

Although we have learned in this chapter about many new exciting features unique to the ferromagnetic Josephson structures, there are still many unanswered questions. Part of even the analytic discussion was phenomenological: see e.g., Eq. (12.8), which was only very imperfectly justified. Many more questions could be asked, such as what are the S and N layer thickness dependencies and even what would be the results of a study of the $\Delta_1 \neq \Delta_2$ Josephson structure. However, what has been done in this chapter leaves, I hope, a good foundation and highlights some of the more important aspects clearly worthy of future study. The theory as given will be useful guidance for future experiments into ferromagnetic Josephson structures and their applications in spintronic devices.

Chapter 13

Concluding Remarks

This is a rather long and at times fairly technical book and I believe it is good to summarize now some concluding thoughts for the benefit of those who made it to the end, or perhaps for those who skipped to it.

The main purpose of the book is to explain how the equilibrium and transport properties of clean Ferromagnet/Superconductor heterostructures can be accurately calculated starting from the standard BCS type Hamiltonian. The main techniques, including numerical methods, are discussed in detail. I have shown in the book that the results that can be obtained are in excellent, detailed, and quantitative agreement with experiment when results obtained in good quality samples exist.

There is ample motivation for wanting to study these structures. The unique oscillatory properties arising from the F/S proximity effect, introduced early in the book, make these heterostructures eminently suitable candidates for many switching devices, ranging from nonvolatile low power memory elements to quantum computing applications involving Josephson junctions. But quite apart from the possible applications, important as they might be, there is a very considerable amount of fundamental Physics involved, and this is what I have tried to emphasize.

A very fundamental phenomenon that kept recurring throughout the book is the existence and importance of singlet to triplet pair conversion: the ordinary s-wave singlet Cooper pairs may convert to any kind of triplet pairs. Breaking the shackles imposed by spin conservation, and even by the Pauli principle, is at the fountainhead of all the amazingly wide variety of results that have been displayed in the many figures that are in the "Results" sections of the different chapters. We tend to think of the restrictions imposed by fundamental symmetry as things that we always have to live with, and the realization that inhomogeneity in the magnetic part of the F/S structure leads to liberation from spin conservation rules and even to what is in effect a way to work around the otherwise immutable Pauli principle is at first very surprising.

One thing I have emphasized in this book is that one must do things right: in evaluating transport properties, for example, conservation laws must be absolutely respected. Self-consistency is not just an esoteric and optional requirement. It is related to charge conservation. This is blissfully ignored in paper after paper through the literature, even though it was pointed out over fifty years ago.

For the student, the techniques that one needs to use (repeated diagonalization of the Bogoliubov-DeGennes equations, transfer matrix methodology, and so on) have been described in very considerable detail. Ultimately, the physics of all of these methods can be easily related to that of analogous problems in first year graduate level Quantum Mechanics. The numerical techniques have been explained in some detail. I have not found it necessary to include any computer codes. This is because every computer system in a research (or even teaching) institution has available the standard matrix diagonalization and matrix manipulation programs that are needed. The only code writing required consists very largely of arranging the proper calls to these "canned" routines and functions, with some attention paid to taking advantage of parallel computing capabilities whenever available. These will save a considerable amount of waiting time.

The book emphasizes throughout clean fabricable nanostructures as far as the composition of the layers, but as for the interfaces, they are assumed to be "good but not perfect" in a completely realistic way. Dirty structures, although they can also be studied quite easily, are of much less interest, and do not have much of a future: as fabrication techniques keep getting better the trend is clearly towards samples of quality even higher than those considered in this book. For application purposes, reproducibility is a must and dirt is always unpredictable.

This is a theory book. But the theory is neither abstruse nor esoteric and, in a sense, not particularly advanced. Knowledge of only introductory graduate physics has been assumed, and even a solid undergraduate level training should be enough. The book can easily be read and understood by experimentalists, and just about anybody can at least look at the pictures and read the explanations that go with them. These pictures are meant to be, overall, a visual guide to what things vary as the values of relevant variables change, and how. What are the trends is what is important.

Perhaps even more important is to be able to spot when is it that there really are not any trends but rather that, because of the more or less intricate oscillatory behaviors, what appears to be a true trend will actually eventually reverse.

My hope here is that the book will prove to be useful and that some day, in a not distant future, I will have the pleasure to see an actual device that was helped into existence by some of the contents of this book. I hope also the book helps its readers understand the role and limitations of symmetries and fundamental principles.

References

[1] G. Deutscher and P.G. de Gennes, *Superconductivity*, R.D. Parks (Ed.) (Marcel Dekker, New York, 1969), p. 1005.

[2] P.G. de Gennes, *Superconductivity of Metals and Alloys* (Addison-Wesley, Reading, MA, 1989).

[3] M. Tinkham, *Introduction to Superconductivity* (Kriegern Malabar, Florida, 1975).

[4] G. Eilenberger, *Z. Phys.* **214**, 195 (1968).

[5] J.B. Ketterson and S.N. Song, *Superconductivity* (Cambridge University Press, Cambridge, UK, 1999), p. 286.

[6] K. Halterman and O.T. Valls, *Phys. Rev. B* **66**, 224516 (2002).

[7] V.L. Berezinskii, *JETP Lett.* **20**, 287, (1974).

[8] K. Halterman, O.T. Valls and P.H. Barsic, *Phys. Rev. B* **77**, 174511 (2008).

[9] I. Kosztin, Š. Kos, M. Stone, and A.J. Leggett, *Phys. Rev. B* **58**, 9365 (1998).

[10] P.H. Barsic, O.T. Valls, and K. Halterman, *Phys. Rev. B* **75**, 104502 (2007).

[11] V.V. Ryazanov, V.A. Oboznov, A.Y. Rusanov, A.V. Veretennikov, A.A. Golubov, and J. Aarts, *Phys. Rev. Lett.* **86**, 2427 (2001).

[12] O. Bourgeois, S.E. Skipetrov, F.Ong, and J. Chaussy, *Phys. Rev. Lett.* **94**, 057007 (2005).

[13] V. Shelukhin, A. Tsukernik, M. Karpovski, Y. Blum, K.B. Efetov, A.F. Volkov, T. Champel, M. Eschrig, T. Löfwander, G. Schön, and A. Palevski, *Phys. Rev. B* **73**, 174506 (2006).

[14] S.M. Frolov, D.J. Van Harlingen, V. V. Bolginov, V.A. Oboznov, and V.V. Ryazanov, *Phys. Rev. B* **74**, 020503(R) (2006).

[15] Y. Obi, M. Ikabe, T. Kubo, and H. Fujimori, *Physica C* **317–318**, 149 (1999).

[16] C.T. Wu, O.T. Valls, and K. Halterman, *Phys. Rev. B* **86**, 014523 (2012).

[17] F. Chiodi *et al.*, *Europhys. Lett.* **101**, 37002 (2013).

[18] C.T. Wu, O.T. Valls, and K. Halterman, *Phys. Rev. B* **86**, 184517 (2012).

[19] C.T. Wu, O.T. Valls, and K. Halterman, *Phys. Rev. Lett.* **108**, 117005 (2012).

[20] K. Halterman, and O.T. Valls, *Phys. Rev. B* **80**, 104502 (2009).

[21] A.A. Jara, C. Safranski, I.N. Krivorotov, C.-T. Wu, A.N. Malmi-Kakkada, O.T. Valls, and K. Halterman, *Phys. Rev. B* **89**, 184502 (2014).

[22] J. Zhu, I.N. Krivorotov, K. Halterman, and O.T. Valls, *Phys. Rev. Lett.* **105**, 207002 (2010).

[23] A.F. Andreev, JETP **19**, 1228 (1964).

[24] D. SaintJames, *J. Phys.* (France) **25**, 289 (1964).

[25] G.E. Blonder, M. Tinkham, and T.M. Klapwijk, *Phys. Rev.* **25**, 4515 (1982).

[26] C. Kittel, *Introduction to Solid State Physics*, 7th ed., (John Wiley and Sons, Inc., New York, NY, 1996).

[27] G. Baym and L.P. Kadanoff, *Phys. Rev.* **124**, 287 (1961).

[28] A. Spuntarelli, P. Pieri, and G.C. Strinati, *Phys. Rep.* **488**, 111 (2010).

[29] C.T. Wu, O.T. Valls, and K. Halterman, *Phys. Rev. B* **90**, 054523 (2014).

[30] E. Moen and O.T. Valls, *Phys. Rev. B* **95**, 054503 (2017).

[31] N.W. Ashcroft and N.D. Mermin, *Solid State Physics* (Philadelphia, PA, 1976). See Appendix C.

[32] E. Moen and O.T. Valls, *Phys. Rev. B* **97**, 174506 (2018).

[33] E. Moen and O.T. Valls, *Phys. Rev. B* **98**, 104512 (2018).

[34] K. Halterman, O.T. Valls, and C.T. Wu, *Phys. Rev. B* **92**, 174516 (2015).

[35] X. Waintal and P.W. Brouer, *Phys. Rev.* **65**, 054407 (2002).

[36] T.P. Orlando and K.A. Delin, *Foundations of Applied Superconductivity* (Addison-Wesley Publishing Company, Reading, MA, 1991) pp. 393–488.

[37] E. Moen and O.T. Valls, *Phys. Rev. B* **101**, 84522 (2020).

Index